RURAL LAND USE ON THE ATLANTIC PERIPHERY OF EUROPE: SCOTLAND & IRELAND

edited by

Alexander Fenton

and

Desmond A. Gillmor

ROYAL IRISH ACADEMY
Dublin 1994

Royal Irish Academy
19 Dawson Street
Dublin 2
Ireland

ISBN 1 874045 09 7

British Library Cataloguing-in-Publication Data.
A catalogue record for this book is available from the British Library.

Typeset by Phototype-Set Ltd. Dublin.
Printed by Colour Books Ltd

CONTENTS

CONTRIBUTORS

JEAN BALFOUR is a farmer, forester and landowner. She was formerly Chairperson of the Countryside Commission for Scotland, a member of the Nature Conservancy Council and Vice-President of the East of Scotland College of Agriculture.

PETER CLINCH is a postgraduate student at Nuffield College, University of Oxford, and was formerly at the Environmental Institute, University College Dublin.

FRANK CONVERY is Heritage Trust Professor of Environmental Studies and Director of the Environmental Institute, University College Dublin.

TERRY COPPOCK is Secretary and Treasurer of the Carnegie Trust for the Universities of Scotland, Dunfermline, and formerly Professor of Geography, University of Edinburgh.

DONALD DAVIDSON is Professor of Environmental Science, University of Stirling.

ALEXANDER FENTON is Professor of Scottish Ethnology and Director of the School of Scottish Studies, University of Edinburgh.

DESMOND GILLMOR is Associate Professor of Geography and Head of Department, Trinity College, University of Dublin.

DAVID HICKIE is Environmental Officer, An Taisce, Dublin.

JOHN LEE is Acting Head, TEAGASC Johnstown Castle Soils and Environmental Research Centre, Wexford.

DEREK LYDDON is a chartered architect and planner in Edinburgh and was formerly Chief Planner in the Scottish Office.

DONALD MACKAY was formerly Under-Secretary in the Scottish Office, responsible for countryside and planning issues.

ALEXANDER MATHER is Senior Lecturer in Geography, University of Aberdeen.

KAY MILTON is Lecturer in Social Anthropology, Queen's University, Belfast.

JOHN POLLARD is Senior Lecturer in Geography, Department of Environmental Studies, University of Ulster at Coleraine.

GUY ROBINSON is Senior Lecturer in Geography, University of Edinburgh.

JAMES WALSH is Senior Lecturer in Geography, St Patrick's College, Maynooth.

INTRODUCTION

This book is the proceedings of the conference 'Rural land use on the Atlantic periphery of Europe: Scotland and Ireland' held at the Royal Irish Academy, Dublin, on 17 and 18 September 1992. The conference was a joint meeting of the Royal Society of Edinburgh and the Royal Irish Academy, the premier learned societies in Scotland and Ireland respectively. The Royal Society of Edinburgh was founded in 1783 and the Royal Irish Academy was incorporated in 1785, but this was the first joint meeting between the two societies. It constituted, in the words of the President of the Royal Irish Academy, Professor Aidan Clarke, "a unique coming together".

Because of the great significance of the occasion, it was decided to make the theme of the meeting one which is of fundamental importance to both Scotland and Ireland. It was felt also that the subject chosen should appeal to a wide range of interests rather than represent a narrow specialism. Both these criteria were fulfilled admirably by having land use as the conference theme. Only about 3% of Scottish and Irish land is urban and the land is almost totally rural outside of the Central Valley in Scotland and the Dublin and Belfast urban regions in Ireland. Thus the conference focused on the use of rural land, but this is not to deny that rural and urban are closely interrelated. Rural land use must be seen also within the broader contexts of national and international economies and societies.

As close neighbours on the Atlantic periphery of Europe, Scotland and Ireland share many common features of their physical environments and historical evolutions. The similarities between the two countries were emphasised by the President of Ireland, Mrs Mary Robinson, on the first official visit of an Irish head of state to Scotland two months before the conference. Similarities in settlement histories, two-way movements of people, and centuries of both being parts of the United Kingdom have contributed to the many cultural, social, political and economic links between Scotland and Ireland. These physical and human similarities affect the use of rural land.

Divergence of administrative influences dates in particular from 1922, with the independence of what was at first the Irish Free State and since 1949 the Republic of Ireland, with Scotland and Northern Ireland remaining within the United Kingdom. Thus a threefold distinction can be made in some respects between Scotland, Northern Ireland and the Republic of Ireland, with Northern Ireland often being in an intermediate position. Treatment of Ireland in this book is of the whole island and that is what is meant when the word Ireland is used. Comparisons between Northern Ireland and the Republic of Ireland are made where appropriate.

Since the United Kingdom and the Republic of Ireland both joined the European Community in 1973, Scotland and Ireland became subject to certain common influences affecting rural land use, most notably the Common Agricultural Policy. Thus a further dimension to the interest of studying land use in the two countries is the unusual situation of two territories which had been within one political unit and in which subsequent administrative divergence has been succeeded by convergence. Both Scotland and Ireland are now on the Atlantic periphery of the European Community, having in common many of the

1

problems and the opportunities which this confers. Knowing more about each other's land use, sharing experiences and discussing policies should be of mutual benefit. Coming at a time when major changes in rural land use are being instigated by the European Community, this investigation is all the more apposite.

The first four papers provide a framework for later discussion of individual land uses. Scotland and Ireland should be seen in the European context, and this overview is provided with regard to land use patterns and trends by Alexander Mather in the first paper. Then John Lee compares the biophysical environments of Ireland and Scotland in terms of their suitability for agricultural and forestry production. In the next two papers Terence Coppock and Desmond Gillmor trace the evolution of rural land uses in the twentieth century in Scotland and Ireland respectively, considering the sources of information on land use and the patterns and trends which they portray.

Thereafter the major emphasis in the book is on the principal individual rural land uses in the two countries — agriculture, which remains the predominant use of rural land, and the increasingly significant uses of forestry, recreation and conservation. These are dealt with in a sequence of paired papers by Scottish and Irish contributors: Guy Robinson and James Walsh on agriculture, Donald Mackay and Frank Convery with Peter Clinch on forestry, Jean Balfour and John Pollard on recreation, and Donald Davidson and David Hickie on conservation. In their treatment of these topics, authors were not required to adhere to a repetitive standardised format. In two integrative concluding papers, Derek Lyddon and Kay Milton consider aspects of rural land-use planning and management. An abstract is provided at the beginning of each paper.

Major requirements which emerge from the papers and which were emphasised in discussion at the conference include the need for much more information on rural land use and its changing patterns, the need for greater coordination between the many organisations and interests involved in rural land use, and the need for evaluation of policies and for improved and comprehensive interrelated policies at European Community, national, regional and local levels. A call which was made strongly in discussions was for much greater input to policy-making from the grassroots level. These issues and the many other aspects of rural land use dealt with in respect of Scotland and Ireland in this book should be of interest to those concerned with rural land use anywhere, and in particular in the context of the European periphery.

Although discussion was based on the two countries, it took full cognisance of major issues affecting the lives of all people at the present time. These issues, such as global warming, pollution in all its forms and depletion of fossil fuels, are in varying degrees emotive. They relate on the micro-scale to individuals and on the macro-scale to the world in general, and therefore are seen as matters that can be used as political weapons. Nevertheless it was felt essential that the individual be kept in focus and that planning, indicative or otherwise, should address itself to solving practical problems on the ground. Scotland and Ireland, as peripheral zones from the viewpoint of Europe, might well lend themselves to a test-bed role in planning that honestly and altruistically takes into account the human factor, and that is seriously working for the future of the people themselves.

There was a strong general feeling at the concluding session of the conference that the momentum should be maintained through the setting up of a continuing working group and the holding of another conference. These should investigate further the similarities and differences in land use between Scotland and Ireland, and endeavour to promote the types of planning and policies that would really reach grassroots level.

The editors wish to express their appreciation to the councils and staff of the Royal Society of Edinburgh and the Royal Irish Academy and in particular to those staff of the Royal Irish Academy who organised the conference and the publication of these proceedings. Professor Patrick Cunningham, Food and Agricultural Organisation, Rome, contributed to the early planning of the conference. The editors are very grateful for the enthusiastic involvement of the contributors, without whom the conference and this book would not have been possible. Sessions at the conference were chaired by Professor Aidan Clarke, President of the Royal Irish Academy; Sir Alastair Currie, President of the Royal Society of Edinburgh; Professor Alexander Fenton, Director, School of Scottish Studies, University of Edinburgh; Dr Pierce Ryan, Director, TEAGASC, Dublin; Professor Ronald Buchanan, Director, Institute of Irish Studies, Queen's University, Belfast; and Professor Bruce Proudfoot, University of St Andrews, who also introduced the concluding discussion.

The Royal Irish Academy and the Royal Society of Edinburgh gratefully acknowledge the financial assistance of the European Union and of Scottish Natural Heritage in meeting part of the costs associated with the conference and the publication of these proceedings.

Alexander Fenton and Desmond A. Gillmor

In: A. Fenton and D.A. Gillmor (eds) 1994 *Rural land use on the Atlantic periphery of Europe: Scotland and Ireland*, 5–24. Dublin. Royal Irish Academy.

EUROPEAN LAND USE: AN OVERVIEW

Alexander S. Mather

Abstract: The remaining years of the twentieth century are likely to be a period of rapid change in land use in Europe, as was the late nineteenth century. An enduring pattern of spatial variation in agricultural intensities is described, but it is suggested that it may be modified in the years ahead. Some of the distinctive features of European land use are identified, and trends of contraction of arable land, expansion of woodland and extension of protected areas are discussed. The relative strengths of these trends, and the interactions between them, may vary in a pattern that is differentiated between core and periphery.

Introduction

Many contrasts in rural land use exist between Europe and the other continents. These contrasts apply especially to the major trends in land use (Table 1). Essentially, those in Europe are the reverse of those of the rest of the world, and particularly of the developing world. These differences stem from environmental conditions, from the long and distinctive history of environmental modification by humans, and from the high population densities and high levels of technology in Europe. As the world economy has expanded spatially and global trading patterns have developed, the demands made on European land have changed. In the past, towns and cities were supplied with food, fibre and wood produced locally. Today, there is a world-wide supply system for these

TABLE 1. Land-use trends (after Richards 1990).

| | *Percentage changes* | | | |
	1700–1850	*1850–1920*	*1920–1950*	*1950–1980*
Europe				
Forests and woodland	−10.9	−2.4	−0.5	6.5
Cropland	97.0	11.4	3.4	−9.9
World				
Forests and woodland	−4.0	−4.8	−5.1	−6.2
Cropland	102.6	70.0	28.1	28.3

5

products, but increasing demands are placed on local land for pleasant residential and recreational environments. Ironically, as farming has become more a business and less a way of life, the countryside has increasingly been perceived by urban-based populations as offering pleasant living conditions. Counterurbanisation and the growth of rural industry have affected many areas.

The history of land use is often characterised by bursts of rapid change interspersed between longer periods of gradual, incremental adjustment. In Europe, the late twentieth century is likely to be a period of rapid change in land use, as was the late nineteenth century. Towards the end of last century, new trading patterns resulting from the opening up of the prairies and other parts of the New World had major effects in Europe: many European farmers simply could not compete with cheap imports of grain and other products. Government responses differed in different countries. In Britain, *laissez-faire* prevailed: land passed out of cereal production in the south and east, and much land in Scotland passed out of sheep-farming and into grouse moors and deer forests. In France and Germany, protection was afforded against cheap imports. In countries such as the Netherlands and Denmark, new forms of specialisation in perishable horticultural and livestock products were encouraged.

A similar period of adjustment is probably under way at present. Over-production of some high-cost agricultural commodities is an expensive and embarrassing problem. The climate of international trade in land products is changing, as it did at the end of the nineteenth century. The environmental consequences of the forms of land use practised in recent decades have attracted increasing opposition, especially from urban-based populations which now greatly outnumber those depending on the land for their living. During the 1980s, signs of land-use change became increasingly well defined across the continent, and they are probably set to continue in the foreseeable future.

These changes will be reviewed, but it should be borne in mind that strong elements of continuity remain, especially in the spatial pattern of land-use activity.

Continuity

In most of Europe, agriculture is the dominant rural land use (Appendix 1 indicates its relative extent). Types of agricultural land use vary, not least in accordance with climate and soils, but the spatial pattern of the intensity of agricultural land use has been remarkably constant over several decades. This pattern is essentially one of contrasting intensities in core and periphery. Figure 1 clearly shows a core–periphery pattern of intensity, based on agricultural yields in the 1930s.

This pattern had become established at least as early as the 1920s, and was related to the pattern of population and to that of technological advances in agriculture. During the eighteenth and nineteenth centuries, the agricultural revolution occurred earliest and most completely around the main centres of industrialisation in western Europe, where both capital and technology were available (Mazoyer 1981). Jonasson (1925–6) viewed northwest Europe as one vast conurbation and one geographical centre of consumption and market facilities for agricultural production. The single-core model visualised by

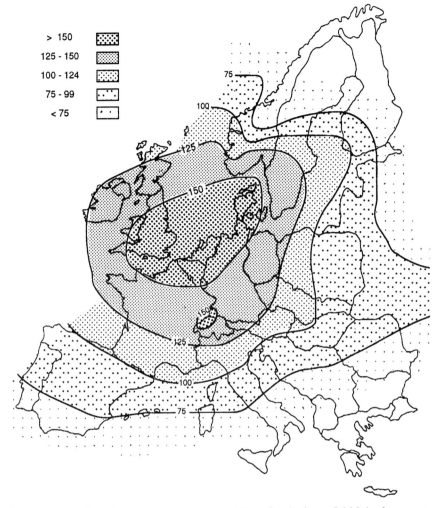

Fig. 1. Intensity of agriculture in Europe, 1930s: the index of 100 is the average European yield of eight main crops (after Van Valkenburg and Held 1952).

Jonasson and portrayed in Fig. 1 is, of course, a generalisation and simplification of reality. At a different scale, several sub-cores or national cores could be recognised, for example around cities such as Dublin, Edinburgh and Oslo. Similarly, national peripheries also exist in addition to the continental periphery: in the cases of Scotland and Ireland these coincide.

Figure 2 and Table 2 confirm that essentially the same continental core–periphery pattern prevails more than half a century later, despite radical changes in political and economic climates, including the demise of overseas empires and the emergence of the European Community (EC) and its Common Agricultural Policy (CAP). The core–periphery pattern is clearly enduring, and will not be easily replaced. If and when it does fade in terms of patterns of agricultural intensity, it may well simply mutate into other forms.

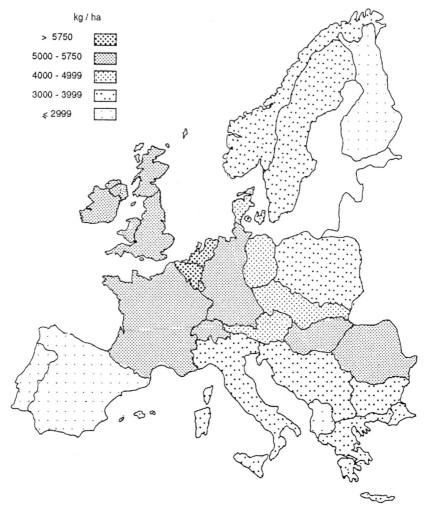

kg / ha

> 5750

5000 - 5750

4000 - 4999

3000 - 3999

≤ 2999

Fig. 2. Cereal yields 1986–88 (after WRI 1990).

Change

Agriculture: arable/cropland trends

While the arable area is expanding in most other continents, it is contracting in much of Europe (Table 1). The combination of rising crop yields and stagnating populations and demand for food has led to surplus production, or at least to surpluses of expensive food that cannot readily be sold on the international market. The result is that substantial areas of arable land are now surplus to requirements, and are passing out of arable use.

In much of Europe during the nineteenth century, population pressures of the kind now associated with the developing world were widespread, especially in the peripheries of the constituent countries. In areas such as the Atlantic fringes of Scotland and Ireland, as well as in the French Alps and the south of Italy, land

TABLE 2. Agriculture: intensity of selected inputs (after OECD 1991b).

| | Fertilisers* | | Pesticides† |
	Nitrogen	Phosphates	
Belgium	12.8	5.7	872.6
Netherlands	21.6	3.8	952.5
Denmark	13.9	3.6	208.4
France	8.5	4.8	290.1
Germany (West)	13.1	5.6	249.6
Greece	4.7	2.1	383.0
Italy	5.4	4.2	1150.1
UK	7.8	3.0	213.5
Ireland	6.2	2.6	39.8
Spain	3.6	1.8	440.7
Portugal	3.5	2.0	286.2

*Tonnes (nutrient content)/km^2.
† Tonnes (active ingredients)/thousand km^2 (area is cropland and permanent pasture combined).

was brought under the spade and plough as the local populations rapidly expanded. Much land of low fertility (but in some cases of high fragility) was pressed into service, as indeed is the case in many parts of the developing world today. With a subsequent drift of population to the cities or overseas, the pressures eased, and gradually arable use was discontinued in many marginal or submarginal areas. More recently, deliberate policy measures have been adopted by the EC to reduce the arable area (or at least the area under certain crops), in the face of over-production and the associated costs of farm support.

The pattern of reduction of the arable area is shown in Fig. 3. Here, as in other sectors, the statistical data need to be viewed with caution. It is clear, however, that most countries have experienced a reduction in extent of arable land, and this reduction is greater in some of the peripheral countries, such as Portugal, Ireland and Finland, than in some core countries, such as Germany and the Netherlands. Reductions in arable areas are not, of course, confined to EC countries: they have occurred widely in eastern and western Europe alike.

There is some evidence that the downward trend in arable extent is accelerating. Table 3 compares the rate of change in arable area in 1989–90 with the average annual rate over the period 1985–90. There is also some (but inconclusive) evidence that the tendency for change is greatest in the peripheral areas. In the light of over-production of cereals in the EC and attempted remedial measures, it is not surprising that the rates of change are greater for the cereals area than for arable land in general.

Whether or not the rate of reduction in arable extent has up to now been greater in the periphery than in the core, it is unlikely to be constant throughout the continent. Contraction may depend partly on the level of compensation payments for 'set-aside' land, but more generally it is likely to be towards the areas of optimal environmental conditions for crops. These correspond to a considerable degree to the continental and national cores. In other words, land

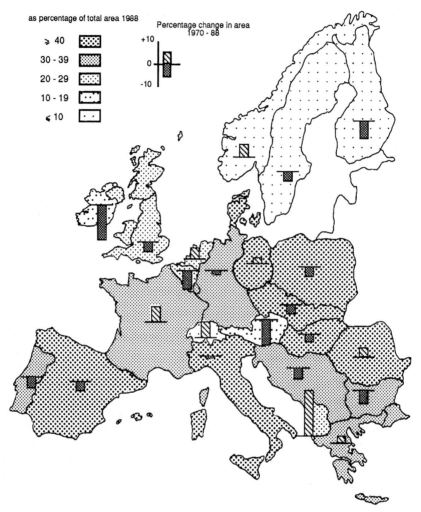

as percentage of total area 1988

Percentage change in area
1970 - 88

≥ 40

30 - 39

20 - 29

10 - 19

≤ 10

+10

0

-10

Fig. 3. Arable extents and trends in arable areas, 1970–1988 (based on data in OECD 1991a for OECD countries and FAO (annual) for eastern Europe).

that is physically, spatially and economically marginal is most likely to pass out of arable use. For many crops, peripheral areas are marginal in all three senses.

The fate of 'surplus' arable land has been the focus of much discussion and speculation in recent years. Estimates of the amount of 'surplus' land in the ten-member EC ranged from 7 to 12 million ha (Conrad 1987; Ilbery 1990). This may be compared to 1.9 million ha under set-aside in 1991 (CEC 1992) and the total land area of Scotland of 7.7 million ha. The total utilised agricultural area has been decreasing for many years, at an average annual rate of 0.4% in the EC (CEC 1987). A 'cascading' process has probably been in operation whereby arable land becomes grassland, grassland becomes rough grazing, and rough grazing passes out of agricultural use — usually either to remain unused or to be afforested.

TABLE 3. Annual (percentage) rates of change, arable land and cereal area,
EC countries (after CEC 1992).

	Arable land		Cereals area	
	1985–90	*1989–90*	*1985–90*	*1989–90*
EC 12	–0.1	–0.6	–1.2	–4.6
Belgium	–0.9	–3.7	–0.6	–4.3
Denmark	–0.5	–0.7	–0.4	0.3
France	0.1	0.5	–1.4	–4.0
Germany (West)	0.1	0.2	–1.8	–3.6
Greece	0.0	0.0	–0.2	1.4
Luxembourg	0.2	–0.2	–1.0	–3.7
Ireland	–1.0	–1.5	–3.9	–5.1
Italy	–0.5	–1.0	–1.7	–3.9
Portugal	0.0	0.0	–4.3	–28.4
Spain	0.0	0.0	–0.1	–4.6
UK	–1.2	–1.2	–1.8	–5.6

Most of the recent and prospective loss of arable land is probably to grassland. Prospects for novel agricultural and industrial crops appear to be limited, and in particular in the view of Ilbery (1990) are unlikely to be bright for smaller-scale farmers in marginal agricultural areas. Prospects for extensive afforestation of surplus arable land are little different. It is true that policies towards the afforestation of arable land have been transformed over the last decade. In countries as different as Britain and Austria, encouragement is now given to the afforestation of arable land, whereas prior to the 1980s it was confined to marginal land. In the case of the latter, for example, the emphasis is being shifted from the afforestation of poor agricultural land to highly productive areas in order to reduce agricultural surpluses (FAO 1988). Prime arable land, however, is still protected (at least in the case of Britain), and in any case its conversion to conventional forest is unlikely to be economically attractive. The likelihood of afforestation of abandoned marginal land may be greater, but even here there may be impediments. In Jutland, for example, where much abandoned land has been afforested over the last hundred years, demands have arisen that instead of being afforested such land should be returned to its natural state and used for recreation (Jensen 1986).

Forest/woodland trends
If one distinctive feature of European land use is a shrinking arable area, another is an expanding area of forest and woodland. Definitions and other statistical difficulties pose even greater problems than in most other sectors of land use, but the overall pattern is clear. Up until the end of the nineteenth century, the forest area shrank in Europe as it is still doing in the developing world today (Table 1). During much of the present century, European forests have expanded. The relative extent of forest and woodland is very variable across Europe, with the forest-poor countries of the UK and Ireland contrasting with

12

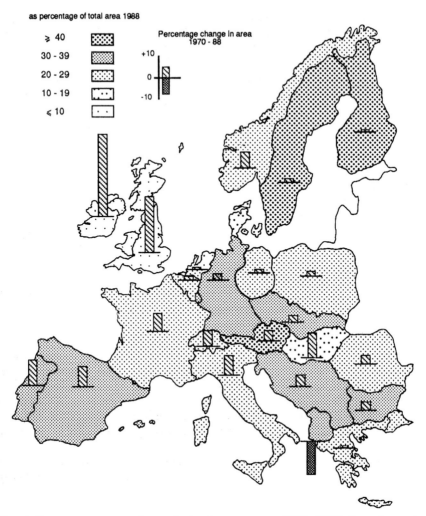

Fig. 4. Forest extents and trends in forest areas, 1970–1988 (based on data in OECD 1991a for OECD countries and FAO (annual) for eastern Europe).

Scandinavia and central Europe, but the area of forest and woodland has expanded in the great majority of European countries (Fig. 4). Once more, there is a tendency for the expansion to be most rapid in some of the peripheral countries. Expansion rates are low in the already well-wooded Scandinavian countries, but are relatively high in Italy and Portugal as well as in Britain and Ireland. Furthermore, there has been a clear tendency for afforestation to be concentrated in the peripheries of these countries, at least until very recently.

The trend of forest expansion is longer established than that of contraction of the arable area. Expansion may have accelerated in some European countries in very recent years, and developments in the CAP would suggest that it will continue to do so in the foreseeable future. Yet afforestation has been occurring for many decades. While deforestation continued in many European countries during the nineteenth century, forests have continued to expand throughout much of the present century, albeit at variable rates. Table 4 illustrates the scale and progress of forest expansion in France, Hungary and Italy.

Two main factors underlie the expansion of forests in European countries. Firstly, stable population and rising crop yields mean that land can be released from agricultural production. This process, combined with the depopulation of some of the mountainous areas and other marginal or harsh environments that had experienced population pressure during the nineteenth century, allowed natural regeneration of forests to occur on abandoned land. This process was already taking place in some of the French mountainous regions, for example, by the end of the nineteenth century (Douguedroit 1981). In other areas, such as parts of Scotland and Ireland, natural regeneration was inhibited by heavy stocking by sheep and deer, even if local seed sources were available. Secondly, many European countries embarked on policies of deliberate forest expansion in the nineteenth or early twentieth century, and have maintained these, with fluctuating degrees of enthusiasm, ever since. The objectives have varied from country to country and have been variable through time — strategic, economic, social and environmental goals have all featured in many national policies — but the result has been sustained and significant expansion. Most of Europe, and especially its Atlantic periphery, is more extensively wooded than 50 years ago, and some parts are probably more extensively wooded than they were several centuries ago.

Compared with Scandinavia and central Europe, much of the Atlantic periphery has environmental conditions that permit rapid growth of major industrial tree species. This is true of spruces and pines in Scotland and Ireland, and of fast-growing eucalypts in Portugal and Spain. Expanding plantations in such areas are one factor attracting new pulp mills, at a time when increased wood production is difficult to achieve in traditional forest-industry countries such as Sweden and Finland (Collins 1992).

The extent to which this expansionary trend of 'industrial' forests in the Atlantic periphery and Iberia may be expected to continue is debatable. On the

TABLE 4. Changing forest areas, France, Hungary and Italy.

France (% of land area)		Hungary* ('000 ha)		Italy ('000 ha)	
1789	14	1800	2765	1910	4654
1862	17	1925	1091	1929	5295
1912	19	1938	1106	1955	5761
1963	21	1950	1166	1965	6089
1970	23	1963	1306	1975	6306
1977	24	1970	1471	1985	6414
1990	27	1980	1610	1990	8063
		1990	1637		

* Present boundaries.
Sources: France – Prieur 1987; Hungary – Keresztesi 1984; Italy – Merlo 1991. Figures for 1990 all from UNEP 1991.

one hand, the availability of surplus farmland, and the fact that the EC is a major net importer of forest products, would point to continued expansion. On the other hand, there are factors that could make major expansion less likely. One potentially significant factor is the even more rapid growth rates that can be achieved in some lower-latitude countries such as Brazil, Chile and New Zealand. With an increasing internationalisation of the forest industry, how will European countries be perceived as locations for investment? Another factor is the opposition that has built up in some European countries (including Britain and Spain) to the continued planting of 'industrial' forests, characterised by uniformity of age and species composition and by primacy of wood production as a management goal. 'Industrial' forests, in the form of new commercial coniferous plantations, have in effect been outlawed from England since 1988. Such plantations are deemed still to be a suitable land use in the British 'periphery', that is Scotland, but even there the forestry industry is coming under increasing pressures and constraints from planning and environmental interests, relating to the location, design and species composition of new plantations.

In contrast to new 'industrial' forests, new 'post-industrial' forests, with higher proportions of broadleaves and more emphasis on recreational and environmental objectives, are generally welcomed in various countries, including Scotland, England, the Netherlands and Denmark. This type of afforestation continues to attract support from environmentalists and other interest groups, whereas the 'industrial' afforestation of previously open land has given rise to bitter controversy in countries as different as Scotland and Finland, especially where it has affected moorlands or wetlands of high conservation value (e.g. NCC 1986; Lehtinen 1991).

Environmental conservation and protected areas

Europe contrasts with most other parts of the world in having expanding forests and shrinking areas of arable land. It resembles most of the rest of the world in having an increasing proportion of its land area under some form of designation for environmental protection or conservation. The significance of conservation designations varies greatly from country to country, both in terms of effectiveness for purposes of conservation and as curbs on land-use change. The effectiveness of designations probably varies between countries and between conservation sectors: perhaps, for example, nature conservation designations have been more effective than those geared to landscape conservation, at least in Scotland. Extensive designations do not necessarily mean effective protection, and international comparisons need to be viewed with great caution. Nevertheless, the rapid expansion of protected areas in much of Europe is clear from Fig. 5.

As yet, no clear spatial pattern of protected areas and their dynamics is evident at the continental scale. Some peripheral countries such as Finland, Ireland and Italy have relatively small extents of protected areas and relatively low rates of increase, but so also have the core countries of Belgium and the Netherlands. On the other hand, the patterns of threatened species of mammals and vascular plants (Fig. 6) are clearer and more familiar: in general, relatively fewer species are 'threatened' in the peripheral countries than in the core. Species may, of course, be 'threatened' by a range of human activities of which agriculture is only one. The similarity of the pattern in Fig. 6 to that of the patterns of agricultural intensity (Figs 1 and 2; Table 2) is, however, noticeable.

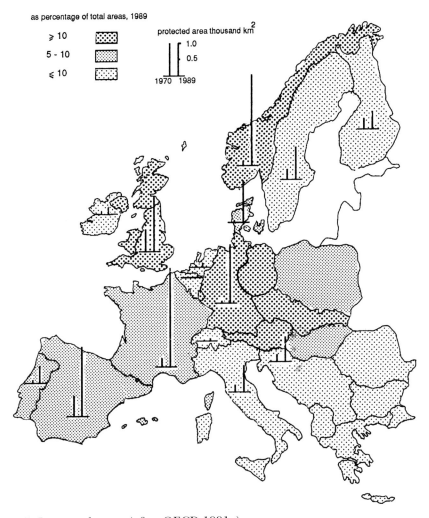

Fig. 5. Protected areas (after OECD 1991a).

At the national scale, protected areas tend to lie in the periphery rather than in the core. The extent of human modification is usually least in the periphery, and the survival of natural and semi-natural habitats greatest. It is to be expected, therefore, that protected areas will be located mainly in the peripheries of individual countries, and perhaps eventually of the continent as a whole.

Such a pattern, though understandable, is in some ways unfortunate. Conservation designations may be seen by local peripheral populations as unwelcome impositions that will inhibit development and the general socio-economic well-being of the remoter areas. Conflict can all too easily result between the rural-dwelling land-users of the periphery and conservation interests whose power bases usually lie in the cities of the core. Such conflict has been severe in the Scottish periphery, where around one-quarter of the area (depending on where the boundaries are drawn) is under some form of designation. It has also

16

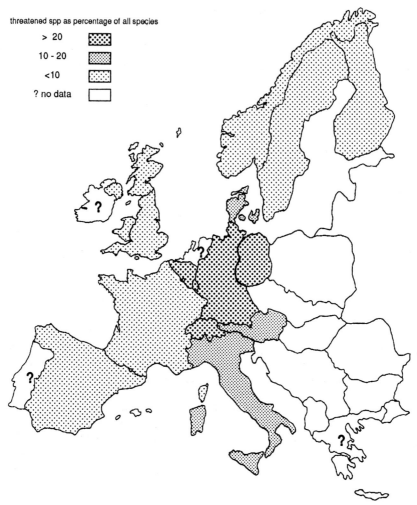

threatened spp as percentage of all species

> 20

10 - 20

<10

? no data

Fig. 6. Threatened species of vascular plants, late 1980s (after OECD 1991a).

occurred in many other areas, including central European locations such as Austria and Hungary (Barker 1982; Persányi 1990): in the South Tirol of Italy, over 43% of the area is designated as conserved landscape and nearly 70% is under some form of protection (Woodruffe 1990). The fact that individuals from the urbanised core are frequently involved in administering and managing protected areas does little to reduce the conflict: the culture gap between them and local populations may be wide (e.g. Dwyer 1991).

The long-established ambivalence of 'core' urbanites' perceptions of the rural periphery does not help. Local residents in the periphery often find it difficult to reconcile the 'core' perceptions that some peripheral locations are suitable for nuclear waste dumps, super-quarries and certain military installations, but that similar locations are of such environmental value that activities such as improved grazing and tree-planting must be prohibited.

Interactions and prospects

Three separate trends in land use in recent years have been briefly reviewed. Agriculture is undergoing retrenchment, while forests and protected areas have been expanding. In considering these trends, three points should be borne in mind.

(a) The trends are not new, although they have accelerated in recent years. As Jansen and Hetsen (1991) point out, agriculture generally has been 'extensified' and has contracted in peripheral areas over the last century, and has been intensified and concentrated in the core areas.

(b) These apparently simple trends reflect net changes. Land-use change is usually extremely complex. Net change is simply the balance between the flow and counter-flow of land between uses, which in turn reflect the decisions taken by numerous land-occupiers in the context of their individual circumstances. Few detailed analyses of patterns of change are available at the national level, and such analyses as are available are often of land cover rather than land use *per se*. The former is more easily monitored, but it is an imperfect indicator of management regimes.

(c) The trends do not occur in isolation but in interaction with each other. The interactions are complex and in some cases difficult to predict. At one level, the role of environmental conservation (in protected areas and in the wider countryside) has grown rapidly, and agricultural and forestry policies are now far 'greener' than even ten years ago. Government incentives for agricultural expansion have almost disappeared, and the rate of habitat loss might therefore be expected to decline. In New Zealand, where the withdrawal of subsidies in the early 1980s was much more complete and abrupt than is likely to occur in Europe, the effects have been viewed as markedly beneficial for habitat conservation (Mark and McSweeney 1990). On the other hand, financial stress in agriculture leads to changes in farm ownership and occupancy, and these in turn often lead to changes in the farming landscape (Munton, Marsden *et al.* 1990).

Sweeping generalisations about the prospects for land use in Europe in the years ahead are unlikely to be helpful or valid. Individual trends and their interactions are likely to vary spatially, and in particular between core and peripheral areas.

Core areas
The long-established pattern of agricultural intensity shown in Fig. 1 reflects the pattern of demand and hence of population. For a number of reasons, we may expect it to undergo change in various ways in the years ahead.

Changing roles of market-location and environmental influences. Compared with the 1920s and 1930s, land transport has increased in flexibility and decreased in relative cost. In addition, free movement across European frontiers is now a reality. The influence of location in relation to the market may therefore be expected to have weakened. Agriculture may thus have been freed, at least to some extent, to seek the optimal environmental conditions for cheap production. It has been observed (see e.g. Belding 1981) that the pattern of intensity illustrated in Fig. 1 is not optimal in relation to 'climatic fertility'. At one scale, therefore, a southwards shift of the zone of peak intensity might be expected. At another scale, a more polycentric pattern of variation in intensity might be expected to develop, reflecting local and national areas of high fertility and of suitability for particular enterprises. In the American South, cropland has become increasingly concentrated on the areas of high land quality (Hart 1978), and increasingly close adjustments between farming and environmental conditions (in particular land capability class) have already been reported from areas as diverse as Canada and New Zealand (Manning and McCuaig 1981; Moran and Mason 1981). In short, a closer adjustment between environmental conditions and agriculture may be expected, and this may have important implications for the continental core. The pattern of potential biomass production is shown in Fig. 7: it is clearly quite different from that of variations in agricultural intensity. The whole question of optimal environmental conditions for crop production is, however, extremely complicated, and is made more so by the possibility of global warming. Such warming is likely to lead to higher yields in northern Europe, and lower yields, because of increased moisture stress, in the south (Parry 1992). It is probably also true that the soils of many of the intensively farmed areas of Europe offer great flexibility of production.

Curbs on high-intensity production. The pattern of spatial variation of fertiliser inputs mirrors that of agricultural intensity and crop yields (Table 2; Figs 1 and 2). Agricultural pollution and intensive methods of production with large applications of fertilisers and other inputs are coming under increasing disfavour. Curbs on intensity of these inputs (and on stocking rates) (CEC 1989) may lead to a flattening of the gradients of intensity depicted in Fig. 1, and to a strengthening of the influence of natural fertility on the pattern of intensive production. Even without such curbs, there may be a trend towards lower-cost, lower-input systems of production.

CARE influences. Centres of high population density may continue to exert a strong influence on land-use patterns, but in a different way from that of recent decades. Instead of functioning as centres of demand for food and fibre, they may act as foci for what McInerney (1986) described as CARE (conservation and rural environment) 'goods'. This new demand is not confined to core or urban areas, but is probably strongest around such areas. Urban regions represent centres of growing demand for rural recreation. Forests are already being established primarily for recreational purposes around some cities in the Netherlands, for example, and there is much talk of 'peri-urban' forests around some British cities. In addition, equestrian enterprises, hobby farming and residential farming (whereby the farm is viewed as a pleasant place of residence rather than as a means of earning a living) are becoming increasingly common around cities, at least in northwest Europe. The long-established intensive agricultural use of land in the core will therefore be exposed to these new demands. Perhaps this trend is not without its benefits: hobby farmers, with their relative independence of income from the farm, may be better conservators of the farmed landscape than those depending on the farm for their living (Munton, Whatmore *et al.* 1989).

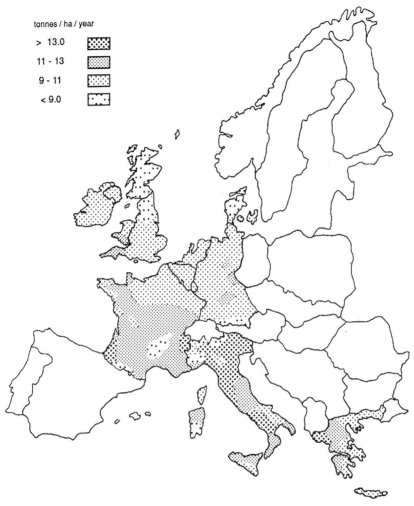

Fig. 7. Simplified pattern of biomass potential, EC 10 (after CEC 1987).

Will the core–periphery pattern of intensity survive? Will the pattern of high intensity of agricultural use around the continental core (and national sub-cores) be modified under the influence of these new circumstances? Will high-intensity food production give way to high-intensity CARE farming? Will the long-established conical pattern of agricultural intensity give way to a pattern more akin to a dough-ring, and will the dough-ring itself be fragmented by the increasing 'pull' of favourable environments and opportunities for low-cost production?

Core areas are often characterised by fertile or prime land, and some conflict may occur between high-intensity agriculture and CARE management. Fertile land, however, does not encircle all core cities: most city hinterlands include hills or other areas of poorer land. Presumably it will be in these areas that CARE land use and management will become most prominent. Conversely, not all areas of highly productive land are located in cores, as Fig. 7 illustrates. Figure 8 suggests

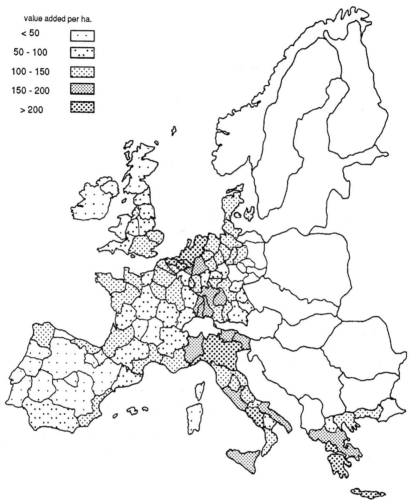

value added per ha.
< 50
50 - 100
100 - 150
150 - 200
> 200

Fig. 8. Simplified pattern of agricultural value added per ha (after CEC 1992).

that there may already have been some movement of intensive production towards some of the environmentally optimal areas. Perhaps tension and conflict may arise in some of these environmentally favoured areas. If particular types of farming become concentrated on them, increasing uniformity of habitat and landscape may be the result.

Peripheral areas

It has been suggested that areas of high conservation value are most likely to occur in the continental and national peripheries. Areas of prime landscape (as opposed to prime land) and of conservation value will presumably continue to be protected, whether in the periphery or elsewhere. In such areas, farmers' incomes may benefit from enterprises based on tourism and recreation as well as from payments of the type offered in Environmentally Sensitive Areas. Land use and landscape in such areas, therefore, may be relatively stable and constant. The question arises, however, as to the extent to which the landscapes of such areas can be frozen in their present forms. The unanswered, and perhaps unanswerable, question is how cultural landscapes can be conserved when the cultures that gave rise to them are decaying. Ironically, the growth of tourism and recreation in areas of high-quality cultural landscapes may simply hasten the decay of the local cultures that produced the now-cherished landscapes.

Intermediate areas

The zone between core and periphery may well turn out to be one of contrasting fortunes. While it may contain some prime land and some prime landscape, much of its area is likely to be undistinguished in terms of both fertility and scenery. At a time of declining agricultural incomes, areas of this type may be especially vulnerable to rapid change in land use and land occupancy. It is in such areas that the greatest likelihood exists of rapid retrenchment of agriculture and of extensive 'industrial' afforestation, if it is going to continue to take place in Europe. The fact that these areas are often relatively remote both from the main urban centres and from the major tourist zones means that the opportunities for farm families to diversify and to become involved in multiple-job-holding may be restricted. Extensive areas of this middle ground exist in both Scotland and Ireland, between the national cores and peripheries and between prime land and prime landscape. The outlook for land use and land users in these areas is uncertain.

A significant factor in this zone (as well as in the core and periphery) is the age structure of farmers. Fewer than one-quarter are under 45, around half are over 55, and many have no identifiable successors (Marsh *et al.* 1991). Most of the over-55 group are likely to leave farming within the next 20 years, whether or not early retirement incentives are offered. The fact that older farmers are likely to dis-intensify their activities may allow a transitional period of gradual change in land use and landscape, but what will happen when they go? Will farm abandonment, and reversion to wildscape, be generally acceptable? Will farm amalgamation and the advent of new owners mean that landscape change accelerates? Will a new group of non-traditional owners replace them, and if so, how will they manage the land?

Review

Radical changes have been made in rural and land-use policies in EC countries in recent years. Even ten years ago, some of the policy changes of the late 1980s and early 1990s would have seemed quite improbable. Perhaps, however, we overestimate the significance of policies and policy changes: it is apparent that essentially similar changes in land use have occurred in recent years under very different political climates in the EC, in other OECD countries, and in the former Soviet-bloc countries.

As we approach the end of the twentieth century, the outlook is for continued and probably accelerating change in land use, perhaps of a scale and magnitude unmatched over the last hundred years. Arable land is likely to continue to pass out of use; the forest area is likely to continue to increase; prime landscapes are likely to be increasingly effectively and extensively protected. Change is unlikely to be evenly distributed across the continent or even across individual countries. It is likely to vary with location and with environmental conditions. In parts of the continental and national cores, agriculture may be expected to dis-intensify, under the influence of growing demands for high environmental quality and for recreation. In other areas of highly productive land, intensive agriculture may be expected to continue and indeed to adjust increasingly closely to variations in environmental conditions. In areas of prime landscape, conservation goals and designations will probably become increasingly prominent. The greatest uncertainty is in the middle ground, where changes in arable extent and forest area are likely to be greatest.

Europe, as has been noticed, is distinctive at the global scale because of its trends in arable and forest land. It is also distinctive in having at present 'surplus' land. The irony is great: the most densely peopled continent has land to spare, while some other continents have desperate shortages of productive farmland. It may be of little consolation to the numerous farming families facing the extinction of their traditional way of life, not least in Scotland and Ireland, but the problems of surpluses are surely preferable to those of shortages. Whether these surpluses (of food and farmland) will remain for long is another matter. In the longer term, the possibility of climatic change and of shortages of fossil fuels and their derivatives should discourage complacency. The temperate and fertile lands of much of Europe may well have new and different demands placed on them within the next few decades. Perhaps, therefore, the usefulness and productivity of these lands should be conserved as well as their beauty and wildlife, and decisions that close off options of increased production of food in the future should be avoided. At a time of over-production and surplus land, and when political horizons are more likely to extend to five years than to fifty years, this may be the greatest challenge facing planners and policy-makers.

References

Barker, M. 1982 Comparison of parks, reserves and landscape protection in three countries of the Eastern Alps. *Environmental Conservation* **9**, 275–86.

Belding, R. 1981 A test of the von Thünen locational model of agricultural land use with accountancy data from the European Economic Community. *Transactions of the Institute of British Geographers* **6**, 176–81.

CEC 1987 *The state of the environment in the EC 1986.* Luxembourg. Commission of the European Communities.

CEC 1989 *Intensive farming and the impact on the environment and rural economy of restrictions on the use of chemical and animal fertilisers.* Brussels/Luxembourg. Commission of the European Communities.

CEC 1992 *The agricultural situation in the Community 1991 report.* Brussels/Luxembourg. Commission of the European Communities.

Collins, L. 1992 Corporate restructuring of the pulp and paper industry in the European Community. *Scottish Geographical Magazine* **108**, 82–91.

Conrad, J. 1987 Alternative land-use options in the European Community. *Land Use Policy* **4**, 229–42.

Douguedroit, A. 1981 Reafforestation in the French Southern Alps. *Mountain Research and Development* **1**, 245–52.

Dwyer, J. 1991 Structural and evolutionary effects upon conservation policy performance: comparing a UK national and French regional park. *Journal of Rural Studies* **7**, 265–75.

FAO (annual) *Production yearbook.* Rome. Food and Agriculture Organisation.

FAO 1988 *Forest policies in Europe.* Rome. Food and Agriculture Organisation.

Hart, J. F. 1978 Cropland concentrations in the South. *Annals of the Association of American Geographers* **68**, 505–17.

Ilbery, B. W. 1990 The challenge of land redundancy. In D. Pinder (ed.), *Western Europe: challenges and change*. London and New York. Belhaven.

Jansen, A. A. and Hetsen, H. 1991 Agriculture and spatial organization in Europe. *Journal of Rural Studies* **7**, 143–51.

Jensen, K. M. 1986 Marginale landbrugsarealer. *Geografisk Tidsskrift* **86**, 69–73.

Jonasson, O. 1925–6 Agricultural regions of Europe. *Economic Geography* **1**, 277–94; **2**, 19–48.

Keresztesi, B. 1984 The development of Hungarian forestry 1950–1980. *Unasylva* **36**, 34–40.

Lehtinen, A. A. 1991 Northern natures: a study of the forest question emerging within the timber-line conflict in Finland. *Fennia* **169**, 57–169.

McInerney, J. 1986 Agricultural policy at the crossroads. *Countryside Planning Yearbook* **7**, 44–75.

Manning, E. W. and McCuaig, J. D. 1981 The loss of Canadian agricultural land: a national perspective. *Ontario Geographer* **18**, 25–45.

Mark, A. F. and McSweeney, G. D. 1990 Patterns of impoverishment in natural communities: case history studies in forest ecosystems — New Zealand. In G. M. Woodwell (ed.), *The earth in transition: patterns and processes of biotic impoverishment*, 151–76. Cambridge University Press.

Marsh, J. *et al.* 1991 *The changing role of the Common Agricultural Policy: the future of farming in Europe*. London and New York. Belhaven.

Mazoyer, M.L. 1981 Origins and mechanisms of reproduction of the regional discrepancies in agricultural development in Europe. *European Review of Agricultural Economics* **8**, 177–96.

Merlo, M. 1991 The effects of late economic development on land use. *Journal of Rural Studies* **7**, 445–57.

Moran, W. and Mason, S. J. 1981 Spatio-temporal localization of New Zealand dairying. *Australian Geographical Studies* **19**, 47–66.

Munton, R. J., Marsden, T. and Whatmore, S. 1990 Technological change in a period of agricultural adjustment. In R. J. Munton *et al.* (eds), *Technological change and the rural environment*, 55–72. London. Fulton.

Munton, R.J., Whatmore, S.J. and Marsden, T.K. 1989 Part-time farming and its implications for the rural landscape: a preliminary analysis. *Environment and Planning A* **21**, 523–56.

NCC 1986 *Nature conservation and afforestation in Britain*. Peterborough. Nature Conservancy Council.

OECD 1991a *Environmental indicators 1991*. Paris. OECD.

OECD 1991b *Environmental data compendium 1991*. Paris. OECD.

Parry, M. 1992 The potential effects of climatic change on agriculture and land use. In F. I. Woodward (ed.), *Ecological consequences of global climatic change*, 63–88. Advances in Ecological Research 22. London. Academic Press.

Persányi, M. 1990 The rural environment in a post-socialist economy: the case of Hungary. In P. Lowe, T. Marsden and S. Whatmore (eds), *Technological change and the rural environment*, 33–52. London. Fulton.

Prieur, M. 1987 Forestry: France. *European Environmental Handbook*, 252–8. London. DocTer.

Richards, J. F. 1990 Land transformation. In B.L. Turner III *et al.* (eds), *The earth as transformed by human action*, 163–78. Cambridge University Press with Clark University.

UNEP (United Nations Environment Programme) 1991 *Environmental data report*. Oxford. Blackwell.

Van Valkenburg, S. and Held, C.C. 1952 *Europe*. London. Chapman and Hall.

Woodruffe, B. J. 1990 Conservation and the rural landscape. In D. Pinder (ed.), *Western Europe: challenges and change*, 258–76. London and New York. Belhaven.

WRI (World Resources Institute) 1990 *World resources 1990–91*. New York and Oxford. Oxford University Press.

Appendix 1

Percentage composition of land use

	Total land (thousand km²)	Cropland	Permanent pasture	Forest and woodland	Other land
EC					
Belgium	33	25	21	21	33
Denmark	42	61	5	12	22
France	550	35	22	27	17
Germany (W)	244	31	18	30	21
Greece	131	30	40	20	10
Ireland	69	14	68	5	13
Italy	294	41	17	23	19
Luxembourg	3	22	27	33	16
Netherlands	34	27	32	9	32
Portugal	92	41	8	32	18
Spain	499	41	20	31	7
UK	242	29	48	10	13
Non–EC					
Austria	83	18	24	39	19
Bulgaria	111	37	18	35	9
Czechoslovakia	125	41	13	37	9
Finland	305	8	—	76	15
Germany (E)	105	47	12	28	13
Hungary	92	57	13	18	11
Iceland	100	—	23	1	76
Norway	307	3	—	27	70
Poland	304	49	13	29	9
Romania	230	46	19	28	7
Sweden	403	7	1	70	22
Switzerland	40	10	40	26	23
Yugoslavia	255	30	25	37	8

Sources: CEC 1992; UNEP 1991.

In: A. Fenton and D.A. Gillmor (eds) 1994 *Rural land use on the Atlantic periphery of Europe: Scotland and Ireland*, 25–38. Dublin. Royal Irish Academy.

A BIOPHYSICAL EVALUATION OF IRELAND AND SCOTLAND FOR CROP PRODUCTION

John Lee

Abstract: Ireland displays a more favourable thermal regime, higher moisture deficits and lower humidity than Scotland. Soil classification shows that the more favourable Cambisol/Luvisol groups are twice as prevalent in Ireland as in Scotland. Whereas Scotland has a predominance of Podzols and Histosols, in Ireland Gleysols and Histosols are dominant. Scotland is also more pronounced topographically, with 38% classified as sloping–steep compared with 21% for Ireland. An assessment of the soils in their agroclimatic context shows that Ireland has a significant advantage in grassland production, with 45% of the area classified in the higher suitability classes A and B compared with 11% in Scotland. Ireland also has a more favourable land base for mechanised arable farming, with 28% classified in the higher suitability classes 1 and 2 compared with 17% for Scotland. An estimated 75% of Scotland is unsuited to arable cultivation, while the comparable Irish figure is 60%. An assessment for forestry potential indicates that 52% of the Republic of Ireland is in yield class 18–24 for Sitka spruce; the comparable figure for Scotland is 28%. However, 40% of Scotland is categorised at yield class 11–21 compared with 20% of the Republic of Ireland.

Introduction

Plant growth and crop production are largely influenced by biophysical conditions related to soil and climatic parameters. Different soils situated in the same climatic environment obviously have different potentials. On the other hand, the same soil occurring in different climatic zones may provide a completely different medium for crop growth. Therefore, despite their obvious impact on growth conditions, soil and climatic parameters can never be considered separately but must be interpreted in an integrated way for proper crop production assessments. The need to integrate soil and climate data for land evaluation purposes has promoted the concept of agro-ecological zonation, which enables soils to be interpreted in their particular agroclimatic context. The main objective of this paper is to evaluate the suitability/productivity of the soil resources of Scotland and Ireland for grassland and arable farming in their agroclimatic context and also to present a comparative assessment of both countries for forestry.

Climatic zonation

Techniques which enable a characterisation of the energy and moisture status of the environment are necessary to permit a definition and characterisation of climatic zones. Such a technique was developed for Scotland (Birse 1971) and later extended to the entire British Isles (Avery 1990). The classification consists of six climatic regime classes based on accumulated temperature and potential soil moisture deficit with altitudinal subclasses (Table 1).

TABLE 1. Climatic regimes in the British Isles (after Avery 1990).

Climatic regime	Mean annual accumulated temperature above 5.6°C (day-degrees)	Average maximum potential soil/moisture deficit (MD) or approximately equivalent potential water deficit (PWD)
Subhumid temperate	> 1375	MD > 125mm (PWD > 100mm)
Humid temperate	> 1375	MD 50–125mm (PWD 25–100mm)
Perhumid temperate	> 1375	MD < 50mm (PWD < 25mm)
Humid (oro[a]) boreal	675–1375	MD 50–125mm (PWD 25–100mm)
Perhumid (oro[a]) boreal	675–1375	MD < 50mm (PWD < 25mm)
(Perhumid) oro-arctic[a]	< 675	MD < 50mm (PWD < 25mm)

[a] Altitude greater than 150m.

The generalised boundaries of the agroclimatic zones for the British Isles are shown in Fig. 1. Scotland is predominantly classified as perhumid boreal or ora-arctic and Ireland is predominantly humid temperate, reflecting in particular a more favourable thermal regime. Ireland experiences higher moisture deficits and lower humidity than Scotland. A more exhaustive classification (W. Verheye, pers. comm.) indicates that the lowland areas of Ireland have a growing degree day value of 1500–2000+ compared with < 1500 in lowland Scotland, thus giving Ireland a clear temperature advantage.

Soils

Soil types

The work of the National Soil Survey of Ireland (J. Diamond, pers. comm., 1992) and the Soil Survey of Scotland (Bibby 1984) provides a basis for comparing the soil resources of both countries (Table 2). For the purposes of establishing a common classification system, the system developed for the production of the 1:1 million soil map of the European Communities (CEC 1985) has been applied.

The soil units are tabulated according to generally accepted principles of soil formation. Fluvisols are developed on material deposited by rivers and are influenced by a flood-plain regime. Gleysols are wet soils dominated by the hydromorphic soil-forming process. Regosols comprise immature soils and show no distinct horizon or layered development. The Rankers are shallow soils developed from silicate rocks in mountainous/hill zones and are characterised

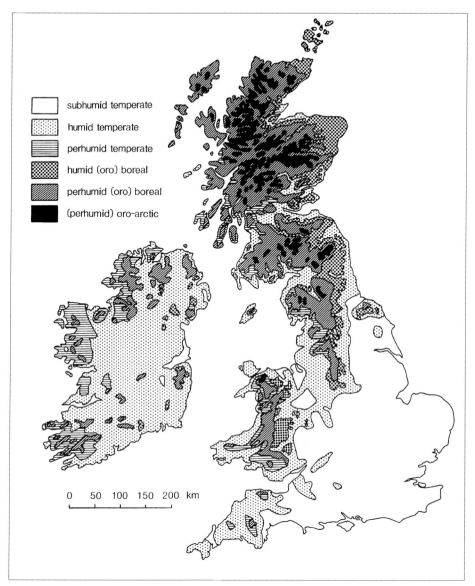

Fig. 1. Schematic map of climatic regimes in the British Isles (after Avery 1990).

by a thin surface horizon. The Cambrisols (Brown Earths) are characterised by weak weathering of rock without significant migration of weathered products within the profile or cross-section. The Luvisols (Grey Brown Podzolics) are characterised by illuviation or washing out of clay under conditions of high base saturation. In the Podzols, illuviation of organic matter and/or sesquioxides is the major determinant. The Histosols (Peats) are very rich in organic matter and are characterised by a thick H horizon (≥ 40–60 cm).

The outstanding features of Table 2 are (i) the predominance of Podzols and

TABLE 2. The major soil units of Scotland and Ireland.

Soil unit	Scotland		Ireland	
	km²	%	km²	%
Fluvisol	950	1	190	0.2
Gleysol[1]	15,290	20	22,270	26
Regosol	500	0.7	780	0.8
Ranker	3,820	5	—	—
Cambisol	11,470	15	15,100	18
Luvisol	—	—	13,065	16
Podzol	15,290	20	13,350	16
Histosol[2]	28,290	37	19,330	23

[1] Includes Brown Earths with gleying.
[2] Includes Peaty Gleys.

TABLE 3. Percentage distribution of slope classes.

	a	a/b	b	b/c	c/d	d
Ireland	7.0	24.5	47.9	20.6	—	—
Scotland	1.3	46.2	14.5	30.6	7.4	—

Histosols in Scotland, (ii) the predominance of Gleysols and Histosols in Ireland, (iii) the absence of Luvisols in Scotland, and (iv) broadly similar proportions of Cambisols in both countries, and indeed broadly similar soil patterns in Ireland and Scotland.

Slope characteristics of land

The slope characteristics of land are significant not only in terms of utilisation but also in terms of susceptibility to water erosion. The legend of the soil map of the European Communities (EC) includes four slope categories as follows:

a = 0–8%	Level	Has no limitations for mechanisation
b = 8–15%	Sloping	Limit of safe use of certain machinery
c = 15–25%	Moderately steep	Has limitations for use of most machinery
d = > 25%	Steep	Unsuited to mechanisation

It is possible to derive from the map legends a slope inventory of the land of Scotland and Ireland (Table 3). It is important to point out that this slope inventory is based on the interpretation of the EC 1:1 million soil map and considers only the dominant slope of each soil map unit. Scotland has a more pronounced topography than Ireland, with 38% classified in the sloping/ moderately steep and steep categories compared with 21% in Ireland. At the other end of the scale 7% of Irish land is classified as level compared with 1% in Scotland.

Fig. 3. Distribution of tillage suitability classes 1 and 2 in Ireland and Scotland.

by our climate. Similarly, Ireland is at a disadvantage geographically in the case of sugar-beet when compared with some of its European partners (Lee 1987). This also applies to the other main arable crops such as potatoes.

Forestry production potential

Republic of Ireland

A provisional land classification according to forestry production potential has been developed for the Republic of Ireland (Bulfin 1991) (Fig. 6). The classification was developed for Sitka spruce, the most commonly planted species, and on the assumption that there are no applications of fertiliser other than those applied at planting. The output of any site is given as a yield class range. Yield class (YC) is the standard method of indicating forestry output and is measured in cubic metres of timber per hectare per annum over the duration of a rotation. Forestry rotations are normally set at the number of years of growth to maximise mean annual production.

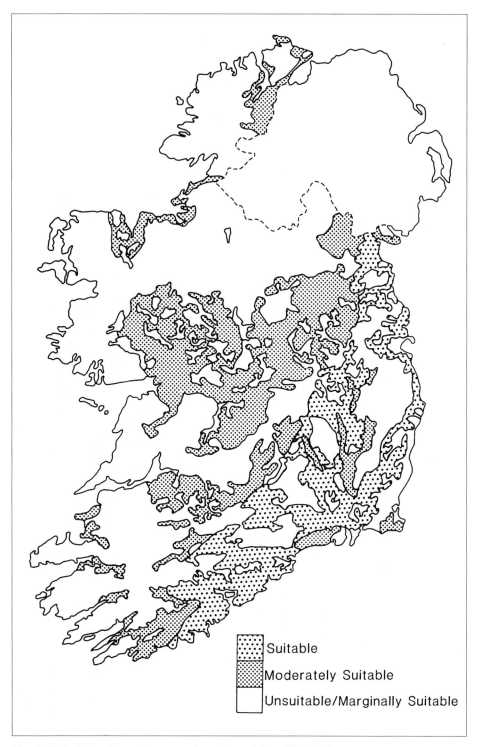

Fig. 4. Suitability for spring cereals — Republic of Ireland.

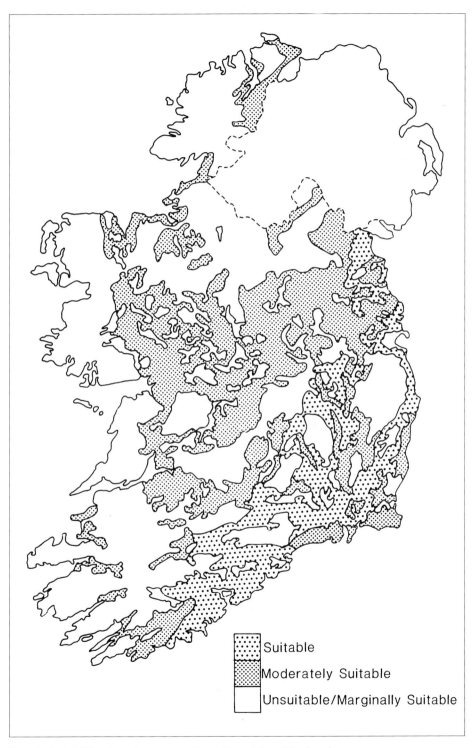

Suitable

Moderately Suitable

Unsuitable/Marginally Suitable

Fig. 5. Suitability for winter cereals — Republic of Ireland.

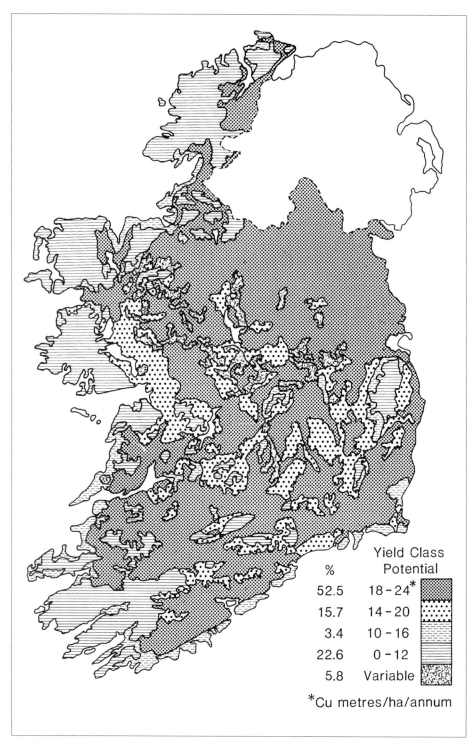

%	Yield Class Potential	
52.5	18 - 24*	
15.7	14 - 20	
3.4	10 - 16	
22.6	0 - 12	
5.8	Variable	

*Cu metres/ha/annum

Fig. 6. Forestry production potential.

There are some very poor sites in the west of the country where other species may be more productive than Sitka spruce. Other species may also be more productive in the east of the country on the more sheltered free-draining soils and in the low-rainfall areas where species such as Douglas fir may yield better (Bulfin 1991). The average production of Sitka spruce within Irish state forests is approximately YC16. Average production in Scandinavia ranges from YC2–6. Average production in the most productive areas of the EC along the Atlantic seaboard would range from YC10 to YC20, with some smaller enclaves giving higher production.

Scotland

The soils of Scotland have been categorised into seven capability classes (Bibby *et al.* 1988) for growth and management of tree crops as follows:

class F1: land with excellent flexibility;
class F2: land with very good flexibility;
class F3: land with good flexibility;
class F4: land with moderate flexibility;
class F5: land with limited flexibility;
class F6: land with very limited flexibility;
class F7: land unsuitable for producing tree crop.

The classification system is based on an assessment of the increasing degree of limitation imposed by the physical features of soil, topography and climate on the growth of trees and silvicultural practices. Species choice is progressively more restrictive from class 1 to class 7. The geographic distribution of the classes is depicted on 1:250,000 scale maps covering the entire country.

The relationship between land capability class for forestry and yield class for Sitka spruce is given in Table 8. The absence of a yield class designation for class F1 land is due to a lack of growth data. For class F7 the yield class designation is biased upwards and applies only to the small pockets of better-quality land in this class (D.C. McMillan, pers. comm.). Effectively class F7 is unsuited to Sitka spruce.

TABLE 8. Relationship between land capability for forestry and yield class for Sitka spruce (after McMillan and Towers 1991).

	Average yield class	95% confidence interval
F1	—	—
F2	22.0	16–28
F3	22.1	17–25
F4	20.0	17–24
F5	17.4	14–21
F6	13.9	11–17
F7	9.9	7–13

TABLE 9. Yield class potential (m³/ha/annum) for Sitka spruce — a comparison of the Republic of Ireland and Scotland.

	Scotland			Republic of Ireland	
Yield class	Yield class range[1]	% area	Yield class	Yield class range	% area
1, 2, 3, 4	17–28	27.7	1	18–24	52.5
5	14–21	17.6	2	14–20	15.7
6	11–17	24.7	3	10–16	3.4
7	7–13	28.1	4	0–12	22.6
			5	Variable	5.8

[1] 95% confidence interval.

Comparison of Scotland with the Republic of Ireland

An attempt to illustrate the comparative productivity of Scotland and the Republic of Ireland for Sitka spruce production on a standardised basis is made in Table 9. The main feature is that whereas 52.5% of the Republic of Ireland is in the high YC18–24, the comparable Scottish figure is 27.7%. However, over 40% of Scotland is categorised at intermediate YC11–21, compared with less than 20% in the Republic of Ireland in YC10–20.

References

Avery, B.W. 1990 *Soils of the British Isles.* Wallingford, CAB International.

Bibby, J. 1984 *National Soils Inventory* (mimeograph). Soil Survey of Scotland. Aberdeen. The Macaulay Institute for Soil Research.

Bibby, J.S., Heslop, R.E.F. and Hartrup, R. 1988 *Land capability classification for forestry in Britain.* Aberdeen. Macaulay Land Use Research Institute.

Birse, E.L. 1971 *Assessment of climatic conditions in Scotland — 3, The bioclimatic sub-regimes.* Soil Survey of Scotland. Aberdeen. The Macaulay Institute for Soil Research.

Brereton, A.J. and O'Keeffe, W.F. 1984 *Climate and grass growth in Europe, a study carried out on behalf of the European Community.* An Foras Talúntais, Johnstown Castle, Wexford.

Bulfin, M. 1991 The right trees in the right places. In C. Mollan and M. Maloney (eds), *The right trees in the right places,* 99–107. Dublin. Royal Dublin Society.

CEC 1985 *Soil map of the European Communities, scale 1:1,000,000.* Luxembourg. Directorate-General for Agriculture, Office of Official Publications of the EC.

Lee, J. 1986 *The impact of technology on the alternative uses for land.* FAST Occasional Paper, EC.DGXII. Brussels.

Lee, J. 1987 European land use and resources, an analysis of future EEC demands. *Land Use Policy* 4 (3), 179–99.

MacMillan, D.C. and Towers, W. 1991 Estimating the yield class of Sitka spruce from the land capability for forestry classification. *Scottish Forestry* 45, 298–301.

MAFF (Ministry of Agriculture, Fisheries and Food) 1988 *Agriculture statistics, United Kingdom, 1986.* London. HMSO.

Turc, L. 1954 Le bilan d'eau des sols; relations entre précipitation, l'évapotranspiration et l'écoulement. *Annales Agronomiques* 5, 491–5.

In: A. Fenton and D.A. Gillmor (eds) 1994 *Rural land use on the Atlantic periphery of Europe: Scotland and Ireland,* 39–53. Dublin. Royal Irish Academy.

SCOTTISH LAND USE IN THE TWENTIETH CENTURY

J. Terry Coppock

Abstract: This paper reviews the main changes in the use of land in Scotland from 1900-1990, with particular reference to the availability of data. Measuring transfers to urban land is one of the least well-documented features, but despite the long-established agricultural census, measuring changes in the extent of agricultural land and of rough grazing is not possible because of changes in coverage. The expansion of land in forestry is both the best-documented and most important change of use. Much of the uplands is in multiple use for a wide variety of purposes, but there is little firm information, either now or in the past.

Introduction

A Rip van Winkle who had been able to view Scotland from a space shuttle on a clear day in 1900 and again in 1990 would have been struck by the broad similarities in the ways in which the land of Scotland was used at these two dates rather than by the differences — the dominance of rough, non-tree vegetation over nearly two-thirds of the country, corresponding broadly with the Southern Uplands and the Highlands; the main expanses of improved farmland in the Central Lowlands, the Tweed Valley and around the Moray and Solway Firths; and the main cities in the central belt, with the outlier of Aberdeen — the picture portrayed at a small scale by the map of land use in the immediate post-war period in the *Atlas of Great Britain and Northern Ireland* (Bickmore and Shaw 1963).

Of course, that would be a very superficial impression of changes in land cover rather than land use, for much of what passes for information on the way in which the land is used is in fact information about what it looks like — in part at least because it is very difficult to establish the former, especially on land on which the level of intensity of use is low, as it is in much of Scotland. This is largely a function of the physical character of Scotland, as described by Lee in this volume — the generally high elevation (by the standards of these islands), the severe climatic constraints, and the thin acid or peaty soils of low fertility that prevail over most of the uplands (Thomas and Coppock 1980). There is a broad correspondence between those stretches of semi-natural vegetation and the land over 250m, although this simple relationship is distorted by the oceanic character of Scotland's climate and the dominance of easterly-moving cyclones, so that rough land reaches to sea-level along the west coast of the mainland and in the islands, while improved land reaches well above the 250m contour on the leeward side of the main upland masses. The climatic tree-line behaves in similar fashion, reaching to over 1000m in the eastern Grampians. These resources have been differently evaluated at different times, depending in part on expected levels of living, perceptions of possibilities and technical advances, but the physical endowment has undoubtedly been the dominant factor in the pattern of land use in Scotland, not least in the past century.

Within this broad structure there have been many changes, both quantitative and qualitative. Some are self-evident to anyone who has observed the Scottish scene over the past fifty years and others are known from the recorded comments of contemporaries; but these are subjective impressions which are extraordinarily difficult to document with any certainty. This is in part due to the nature of the changes themselves, many of which are on land used extensively in various combinations of multiple use; but much is due to the lack of good records (Coppock 1978). It is only now, in the early 1990s, that this issue is beginning to be tackled. Although the availability of sound information on the way in which the land of Scotland is used may seem to be of marginal importance, it is in fact essential to drawing valid lessons from the past and to formulating realistic policies for the future.

The concern of this paper is the use of land rather than the land itself, on which information is more readily available through the work of the soil and geological surveys. In respect of climate, that other essential determinant of rural land use, it is important to note that there are few climatic stations above 200m so that much of what is said about climatic conditions in the uplands is based on extrapolation from scanty data. This becomes of particular importance in any consideration of climatic change and its possible implications for the future use of land in Scotland. For present purposes, too, the question of whether climatic conditions have been constant throughout these ninety years should also be borne in mind, the more so because of the sensitivity to change in these marginal conditions.

Urban land

It might be expected that the most obvious and permanent of changes affecting rural Scotland, the extension of the built-up area, would be well documented, but this is far from being the case. It is only since 1985 that regular data on the transfer of land from other uses to urban development have begun to be collected (Scottish Development Department 1986), although information — of uncertain value and certainly not capable of desegregation to individual cities — has been collected since 1951 as a by-product of the agricultural census, on transfers of cultivated land to urban development and, since 1960, of transfers of agricultural land (Best 1981). These show an average annual rate of transfer of some 1000ha since 1951 (Scottish Office Agriculture and Fisheries Department, annually) (Fig. 1). Much smaller areas will have been transferred from rough grazings or woodland.

The size of the built-up area in 1900 is not known, although there are some data for the Scottish burghs in 1908 and extrapolation back on the basis of estimates for the 1950s and data from the population censuses suggests a figure of c. 90,000ha for 1901, although this procedure ignores changes in standards of

Urban Area 1990-1992

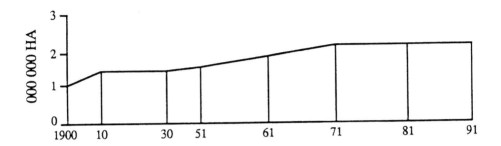

5 Year Average Loss From
Farming to Urban Uses
1945/50-1985/90

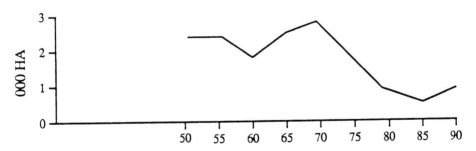

Fig. 1. (a) The urban area of Scotland, 1900–90.
(b) Average transfers from farming to urban uses, from 1945–50 to
1985–90.

Sources: Best 1981; Scottish Office Agriculture Department.

provision, both public and private (Best 1981). The Land Utilisation Survey in
the 1930s was the first attempt to measure the urban area of Scotland directly
and, while the way in which this was done and the fact that much of the
information was collected by schoolchildren cannot have produced data with a
high degree of accuracy, they are the best available and show a built-up area in

Scotland of some 147,000ha, or approximately 2% of the land area of Scotland (Stamp 1948). Twenty years later, Best (1957) attempted to measure the urban area of Great Britain, using a combination of development plans, Ordnance Survey maps and other sources, to give a figure of *c.*190,000ha for Scotland in 1951. This figure includes land occupied by roads and railways beyond the cities, as well as by small settlements and isolated dwellings in the countryside. By calculating provision of land per thousand of population he was also able to provide estimates for both earlier and later years (Best 1981).

It is, of course, common knowledge that the urban area has expanded greatly, with rising housing standards, new factories, airfields and motorways, and five new towns. Figure 1 provides a visual indication of the scale of change, expressed both cumulatively in terms of area occupied and, since 1951, in terms of rate of change. Both sets of figures must be treated with caution; moreover, while the urban area has more than doubled in size, this represents a rise from under 2% to only 3% of the land area of Scotland.

Agricultural land

It might be thought that information on changes in agricultural land would be readily available. After all, there has been an annual agricultural census recording information on the area occupied by different crops and kinds of grassland since 1866 and by rough grazing since 1892 (Coppock 1976; 1978). But the censuses were never intended to provide a record of land use as such and are derived from postal returns from occupiers of agricultural land, or rather from those known to the census authorities. By the beginning of this century the records were probably fairly complete, so that the complication of apparent trends arising merely from more complete enumeration (a serious concern in the first thirty years) was no longer a major cause of distortion. This source is capable of providing a great deal of information about agricultural land — on the crops grown and the balance between the different categories of livestock. However, the emphasis in this paper is on the broad categories of land use, each of which is discussed in more detail elsewhere in this volume.

Improved land

It is probable that changes in the area recorded as improved agricultural land, i.e. that under crops and grass, provide a fairly accurate indication of trends in such land, the source of most transfers to urban development. The area so recorded reached a maximum in the 1900s and then declined fairly steadily thereafter, showing a total fall of *c.*26,000ha between 1900 and 1990, although this is unlikely to be the true figure because of inherent weaknesses in the census as a source of land-use data and the difficulty of keeping track of changes of ownership and occupation, particularly on the urban fringe (Fig. 2).

It is also helpful to draw attention to the main change within the improved land, that between tillage and grassland, not least because tillage is the one relatively unambiguous category which is easy to define on a consistent basis, unlike grassland, which inevitably grades into rough grazing and presents problems of securing uniform interpretation both over time and throughout the country. The area devoted to tillage has experienced greater fluctuations, particularly during the two world wars, when there were major increases at the

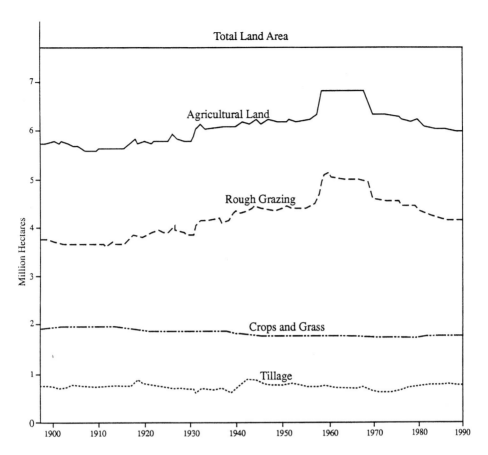

Fig. 2. Agricultural land in Scotland, 1900–90.

Sources: Ministry of Agriculture *et al.* 1968; Scottish Office Agriculture Department.

expense of grassland. On both occasions there was an expansion of crop production westward and upwards, and a subsequent retreat eastward which can be monitored through the agricultural census with some fair degree of certainty, one of the few changes in land use that can be so identified (Coppock 1976).

Rough grazing

It is in respect of changes in the unimproved land used for rough grazing, which accounts for nearly three-quarters of the land used for agriculture, that interpretation is much less certain. The area of such land was recorded for the first time in 1892 and it was a general experience with the agricultural census that several years could elapse after the introduction of a new item before reliable figures began to emerge. In the case of rough land, there was the added complication that its true extent was not of great importance to individual farmers, even if it was known to them, and this was probably even more true of the area actually grazed — there is at least one documented case where land on a hill in southern Scotland was claimed by none of the surrounding farmers (Hart

1955). There is also some evidence to suggest that what farmers often returned was a residual figure determined by deducting other known uses from the total area of the holding, which might itself not be known very accurately.

A further complication was the existence of common grazings in the crofting counties of the Highlands and islands (Moisley 1962), and although there are indications that these have been included in the returns since 1892, the figures had to be revised upwards on several occasions; there were also discrepancies between what the returns recorded and what the secretaries of grazing committees returned (Ministry of Agriculture et al. 1968).

Deer forests were a more serious source of error, although the two issues were related since some common grazings lay in deer forests. Some of the latter, notably the high tops above 1000m, were climatically so extreme and with such poor soils that it is probable that no agricultural use of any kind was made of them — Stamp (1948) estimated that there were some 600,000ha of totally unproductive land in Scotland, most of it in such locations. Originally deer forests were not considered in the censuses, except in so far as they included land that was returned as grazing, but in 1932 the area of deer forests that was either grazed or capable of being grazed was sought and this change led to an increase of c. 35,000ha in the area recorded as rough grazing. In 1959 the instructions were changed again, requiring a return of the whole extent of deer forests, whether they were grazed or not (Ministry of Agriculture et al. 1968). Not surprisingly, there was another large jump of 59,000ha in the total area recorded. A further change occurred in 1970, when smaller agricultural businesses were excluded from the census, leading this time to a fall of c. 37,000ha in the area returned as rough grazing (Coppock 1978).

As Fig. 2 shows, these changes in coverage make it impossible to monitor changes in either the total area of agricultural land or that under rough grazing during the period under review. The latter probably expanded at the expense of improved land in the years preceding the Second World War, when the process was reversed through draining, liming and reseeding, mainly under the stimulus of grants of various kinds. These were continued after the war; but while the fact that such improvements took place is known, neither their extent nor their location can be established from the agricultural census. Nor can the census throw light on the extent or location of any withdrawal of poorer land from agricultural use, although it seems probable that the decline in the market for mutton may have led to some abandonment of the higher land to which the hardier wethers were better suited.

Transfers of agricultural land

The principal destination of land transferred from agriculture, particularly from the 1920s, was to land intended for forestry, and some indication of the scale of change is again provided by the change-of-use data since 1960 and by the land-use data collected by the Ordnance Survey since 1985. Such transfers, which were overwhelmingly from rough grazings (at least in respect of transfers to the Forestry Commission and to private afforestation supported by grant, accounting for the great majority of transfers), averaged some 20,000ha a year since 1960 according to the Scottish Office Agriculture Department. It is, however, not possible to say how much land ceased to be grazed because it was no longer economic to do so, and so joined the category of unused land. As well as the transfers of land to urban development, noted above, land was also taken for recreation and for mineral working, the former averaging some 1000ha annually in the 1980s, though with great fluctuations from year to year, and the latter some 200ha. The annual average recorded loss from agriculture over these forty years was some 20,000ha (Scottish Office Agriculture and Fisheries Department, annually) .

Forestry

While it is known that there has been a considerable expansion in the use of land for forestry, it is impossible to document the extent and location of changes over the past ninety years with any precision. In the early years of the century, data on the area under woodland were collected through the mechanism of the agricultural census, using the information recorded in the parish rate books. But these were not particularly accurate or complete and did not offer a sound basis for monitoring change. The general opinion, from such sources as the Royal Commission on Coast Erosion and Afforestation (1908), was that woodland was neglected and declining in extent, a process accentuated by the demand for home-grown timber during the blockade of the First World War. It was, of course, this latter experience that led the Acland Committee (Reconstruction Committee 1918) to recommend the establishment of a Forestry Commission, and this was done in 1919, with the dual aims of creating a national reserve of timber and of providing help and guidance to private forestry.

The progress of acquisitions by the Forestry Commission, whether of land already under trees or of rough land for afforestation, can be charted with reasonable confidence by reference to its annual reports, although there have been changes in what is recorded that make the examination of long-term trends less certain, particularly in respect of the location of change. As a matter of policy, and initially primarily as a matter of limited finance, the Commission has always sought to acquire unimproved land although, since it is investing public money and seeking to establish productive woodland, it has generally aimed to secure the better unimproved land. It has often had to acquire both improved and unplantable land as part of the process of building up its holdings of woodland, but has attempted to dispose of the former after acquisition. There was inevitably some delay in planting trees on the plantable land and a similar delay may affect the records of the agricultural census of land transferred from agriculture, some of which may never be planted. The great majority of planting in the private sector has been undertaken with the aid of grants under a variety of schemes, and the Commission keeps records of such planting; and although planting may also occur on private land without benefit of grant, the Commission is aware of many of these changes and provides estimates of the scale of such planting in its annual reports.

While the predominant direction of change since 1918 has been an increasing proportion of the land surface of the country under trees, it must not be assumed that the movement is wholly one-way. Some woodland is lost to other uses and the forest census of 1980–2 noted some 16,000ha of woodland which were recorded as such on the most recent Ordnance Survey maps but which were no longer woodland at the time of the census (Forestry Commission 1983). Part of the change may be due to differences in procedure, with the Ordnance Survey continuing to record as woodland areas where the trees have been clear-felled and there is believed to be a prospect of replanting, whereas the Commission will cease to record as woodland cut-over land which has been covered with secondary vegetation. There are also problems over the identification of scrub woodland and differences of opinion about whether to record land with scattered trees as woodland or rough land.

Although some categories of change to and from forestry may escape enumeration, it is nevertheless possible to monitor total changes in the area under woodland since 1918 with a fair degree of certainty (Fig. 3). The formation of the Commission also makes it easier to describe the situation at any one time on the basis of periodic censuses. The first census was undertaken between 1921 and 1926, and is not thought to have been very reliable, although it records the situation after the extensive fellings of woodland during the First World War, with the total area estimated at 435,000ha (Forestry Commission 1928). A second census begun in 1938 was never completed, and the first effective census was that undertaken in 1947–9 (Forestry Commission 1952). It is also the only complete enumeration, at least of woodlands over 2ha (a separate census was conducted of smaller woodland and hedgerow timber), recording this information in great detail. This census similarly recorded the impact of the Second World War, particularly on the private woodlands (the Commission's own woods being mostly too young to be affected), large tracts of which were clear-felled or devastated. The census of 1965–7 was a sample census (Forestry Commission 1970), an approach which necessarily restricts the areal breakdown of the results, as was the most recent census, that conducted in 1980–2, which was confined to private woodland not managed under one of the Commission's schemes of grant aid, information on the Commission's own woods and those private woods that were grant-aided being obtained from the Commission's own records (Forestry Commission 1983). These differences do affect the comparability of the results spatially, although they are adequate for national comparisons. Figures from these last three censuses show an increase of 142,000ha between 1947–9 and 1965–7, and of 84,000ha between the latter and 1980-2, largely through afforestation with conifers on rough land (Fig. 3).

The maps of the Land Utilisation Survey also provide an overview of the extent of woodland, with most of the maps being surveyed in the early 1930s, although they suffer from the weakness that woodland not already shown on the base maps had to be sketched in, often by unskilled observers. They recorded an area of 443,000ha, a figure that is consistent, though not strictly comparable, with those provided by the censuses (Stamp 1948). The nearest contemporary view is that provided by the most recent edition of the Ordnance Survey's 1:50,000 series maps, which incorporate all known large-scale afforestation. These were in fact used in the 1980–1 census and were found to be a reasonably accurate representation of the extent of woodland at that time in Scotland (Forestry Commission 1983).

These records, whatever their limitations for any detailed analysis, show the very considerable increase in the extent of woodland cover of Scotland between 1900 and the present (Fig. 3). In 1900 there may have been some 350,000ha, with the major concentration around the upland margins, particularly in the northeast. To what extent that was reduced before the outbreak of war in 1914 is uncertain, although there is unlikely to have been much new planting/replanting. Felling during the First World War may have totalled some 60,000ha,

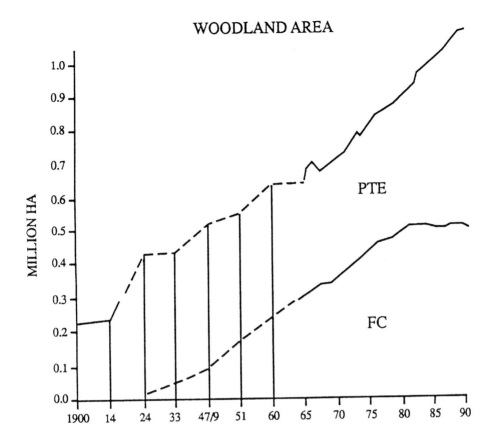

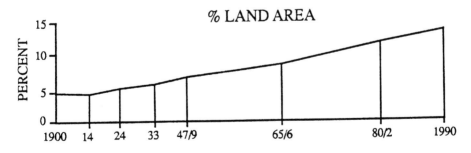

Fig. 3. (a) Woodland area in Scotland, 1919–90.
         (b) Percentage of the land area under woodland in Scotland, 1900–90.

FC = Forestry Commission; Pte = private. Sources: Forestry Commission annually and 1928, 1952 and 1983.

some of which had not been replanted by the census of 1947–9 (Forestry Commission 1928; 1952). The Commission was formed in part because it was felt that the private sector would be unable to undertake any large-scale planting, a view borne out in the inter-war years, when most of the increase was due to the Commission's planting of land that had not previously borne trees in recent times. The scale of such planting was, however, much smaller than the Acland Committee had envisaged, largely because of the limited resources available to the Commission.

New targets for planting were proposed for both the Commission and the private sector in a review document *Post-war forest policy* (Forestry Commission 1943), although these were never accepted by post-war governments, which preferred to set planting targets for five years at a time. Although post-war planting by the Commission has not lived up to the target envisaged in *Post-war forest policy*, it has been on a much greater scale than in pre-war years, involving both replanting of felled and devastated woods acquired from private owners of woodland and new planting, almost entirely on former rough grazings. Private planting has also been on a much greater scale and similarly has included both replanting and new planting, both encouraged by the availability of grants and the latter by tax concessions that encouraged investment from those outside the rural economy. In the past decade, governments with a concern to privatise economic activities in the public sector have reduced the scale of planting by the Commission and required it to dispose of some of its holdings, although this has not yet occurred on a large scale and would not in any case affect the total area under woodland or its distribution. Removal of the tax benefits in the 1988 budget and the recent recession have, however, led to a sharp reduction in the scale of private afforestation.

The main effect of these changes has been greatly to increase the total area of woodland, which now exceeds 1,030,000ha, and to sharpen regional contrasts (Fig. 3). Much of the planting has taken place around the upland margins in the west, southwest and south of the country, where some districts now have more than a fifth of their land under trees, compared with 13% for the country as a whole, a figure which itself contrasts with under 5% in 1900 and 6% in the 1930s.

The emphasis in this account has been very much on the area occupied by woodland, but this is a description of land cover, not of land use. Although most woodland (and certainly that owned by the Commission) is intended to produce marketable timber, other woodland serves primarily an amenity or a recreational purpose (whether for private or public use), and much woodland in these categories is in multiple use. Unfortunately, no sources of information exist to enable estimates to be made of such use, whether in the present or in the past, let alone of trends over time. Shooting is likely to have been an important private recreational use, especially on the larger estates. Policy in respect of woods owned or rented by the Commission has increasingly encouraged multiple use, whether for public recreation, visual amenity or wildlife conservation, and these are increasingly concerns in the giving of assistance to owners of private wood-lands as a condition of grant aid. Since the great majority of private planting since 1919 has been with such aid, it can be said that, at least since 1965, when conservation and recreation became objects of forest policy, multiple use of forest land has been an increasingly important feature.

Over the period under review it seems likely that the area under trees in

Scotland has almost tripled, and it has certainly doubled since 1947–9. For most of the period since 1919, the main agent responsible for that increase has been the Forestry Commission, the role of the private sector being largely concentrated on replanting. From the 1960s, however, an increasing component of new planting has been undertaken by forestry investment companies, acting as agents for private investors wishing to take advantages of tax benefits, at least until 1988. Until the 1980s, the bulk of new planting was undertaken by the Commission, but since then, reflecting a change of government policy, the private sector has been responsible for the larger share (Fig. 3). Whatever the balance between public and private, there is no doubt that, as a result of government policies for forestry since 1919, large tracts of the better uplands, formerly used primarily for the grazing of sheep, are now under plantations of conifers.

Land in multiple use

Although multiple use is a feature of much woodland, most land in multiple use is in the uplands on land covered with semi-natural non-tree vegetation, much of which is used primarily for rough grazing. Among the many other uses which such land also serves are grouse-shooting, deer-stalking, wildlife conservation and a wide variety of public recreational activities, as well as water supply and military training. Unfortunately, little is known about the extent of land used for such purposes and even less about any trends over the period under review. No mechanism exists for the collection of such data, nor can information easily be obtained by observation owing to the low intensity and often intermittent or seasonal character of such uses. There is also considerable overlap between categories. It seems likely, however, that the great majority of rough land is in some kind of multiple use at some time of year.

Perhaps 1.2–1.6 million ha are used for grouse-shooting, usually in combination with sheep-grazing, although the balance between the two varies from estate to estate (Dargie and Briggs 1991). Such use depends on the availability of heather, which must be managed by regular burning and occurs mainly in the eastern Grampians (Tivy 1973). Whether there is now more or less land of this kind than in 1900 is unknown, although there is some evidence that the number of birds shot has declined over the period under review.

Similar uncertainty prevails over the extent of land used for deer-stalking, which is primarily a feature of the higher parts of the central, northern and western Highlands (Clutton-Brock and Albion 1989). Deer-stalking corresponds to some degree with land in deer forests, which seem to have declined in extent over the period under review, having reached a peak at the end of the last century (Dargie and Briggs 1991). However, deer are thought to range over an area twice as large, i.e. 2.8 million compared with 1.4 million ha. Unlike grouse moors, much of the area in deer forests is too high and bleak to be used for any other purpose, although, as noted earlier, there is uncertainty about the extent of withdrawal of sheep-grazing from higher ground.

The extent of land used primarily for the conservation of wildlife, mainly in association with grazing by livestock, corresponds with that in National Nature Reserves, local nature reserves and reserves owned by such bodies as the Scottish Wildlife Trust and the Royal Society for the Protection of Birds, and totalled some

agencies to recognise that information should be a corporate resource, to be collected to common standards and common (or at least compatible) areas. There are two particularly important initiatives. First, the English and Scottish Environment Departments have commissioned the Ordnance Survey to collect information on land-use changes as a by-product of the continuous revision of its large-scale maps. There are problems about this process (which is interesting as an example of one government agency collecting information on behalf of another) in that the changes are those since the map was last surveyed. It also has a strong urban bias and is particularly weak on those aspects of land-use change, relating to the less intensive uses, on which it is most difficult to get reliable information. None the less, it is a start (Scottish Development Department 1986).

The second initiative is the flying of complete air cover of Scotland in 1988–9 and the digitisation of that photography to provide a comprehensive data base of land cover, the first time that this has been done explicitly for this purpose (cover was flown in the late 1940s but not with such a record in mind, and exists simply as a collection of photographs of not very good quality). The great bulk of the cover was flown during the summer of 1988, and hence represents a unique snapshot of the face of Scotland, so that at least scholars in the future will have a baseline from which to measure subsequent changes. Interpretation and digitisation are complete (Macaulay Land Use Research Institute 1992), but this material will provide only a partial source and will need to be related to other data in compatible form collected by other agencies if its full potential is to be realised as part of a geographical information system which can serve the needs of a wide variety of agencies.

References

This paper is based on more than forty years of experience in evaluating sources of data on land use and employing them to examine the changing geography of rural land use in Great Britain (Best and Coppock 1962). Although relating to the situation in the mid-1970s, the author's review of sources (Coppock 1978) spells out in much greater detail than is possible here the limitations of these data and the difficulties of making comparisons over time; this has since been updated in more abbreviated form (Coppock 1991; Coppock and Kirby 1987). The minutes of evidence and appendices of the inquiry by the Select Committee on Scottish Affairs (1972) are a mine of information. What is said about the use of land for agriculture is derived mainly from the Scottish Office Agriculture and Fisheries Department's annual volume *Agricultural statistics Scotland*, summaries of which to 1966 are given in the centenary volume (Ministry of Agriculture *et al.* 1968). Forestry statistics are published in the annual reports of the Forestry Commission and in the reports of successive censuses (Forestry Commission 1928; 1952; 1970; 1983). Both agricultural and forestry statistics are summarised in the annual *Scottish abstract of statistics*. Information on urban land draws heavily on the work of Best (1957; 1981) and that on multiple use on a wide variety of sources; the present position is conveniently summarised in *The Scottish Environment*, an annual publication since 1990 of the Scottish Office.

Best, R.H. 1957 The urban area of Great Britain — an estimate of the extent of urban land in 1950. *Town Planning Review* **28**, 191–208.

Best, R.H. 1981 *Land use and living space.* London. Methuen.

Best, R.H. and Coppock, J.T. 1962 *The changing use of land in Britain.* London. Faber.

Bickmore, D.P. and Shaw, M.A. (eds) 1963 *Atlas of Great Britain and Northern Ireland.* Oxford. Clarendon Press.

Clutton-Brock, T.H. and Albion, S.D. 1989 *Red deer in the Highlands.* Oxford. Blackwell.

Coppock, J.T. 1966 The recreational use of land and water in rural Britain. *Tijdschrift voor Economische en Sociale Geografie* **57**, 81–96.

Coppock, J.T. 1976 *An agricultural atlas of Scotland.* Edinburgh. John Donald.

Coppock, J.T. 1978 Land use. In W.F. Maunder (ed.), *Reviews of United Kingdom statistical sources,* **8**. Oxford. Pergamon.

Coppock, J.T. 1991 Rural land use. In M.J. Healey (ed.), *Economic activity and land use: the changing information base for local and regional studies,* 81–9. London. Longman.

Coppock, J.T. and Kirby, R.P. 1987 *Reviews of approaches and sources for monitoring change in the landscape of Scotland.* Edinburgh. Scottish Development Department.

Dargie, T.C.D. and Briggs, D.J. 1991 *State of the Scottish environment.* Report to Scottish Wildlife and Countryside Link, Perth.

Forestry Commission, annual reports.

Forestry Commission 1928 *Report on census of woodlands and census of production of home growth timber, 1924.* London. HMSO.

Forestry Commission 1943 *Post-war forest policy.* Cmd. 6477. London. HMSO.

Forestry Commission 1952 *Census of woodlands 1947-1949.* Census Report No. 1. London. HMSO.

Forestry Commission 1970 *Census of woodlands 1965–67.* London. HMSO.

Forestry Commission 1983 *Census of woodlands and trees 1980–82.* Edinburgh. HMSO.

Hart, J.F. 1955 *The British moorlands.* University of Georgia Monographs, No. 2. Athens, Georgia.

Hendry, G.F. 1958 The size of Scotland's bracken problem. *Scottish Agricultural Economics* **9** (2), 21–8.

Macaulay Land Use Research Institute 1992 *An analysis of methodology used to measure land cover change with special reference to an evaluation of a GIS performance.* Aberdeen. Macaulay Land Use Research Institute.

Ministry of Agriculture, Fisheries and Food, Department of Agriculture and Fisheries for Scotland 1968 *A century of agricultural statistics. Great Britain 1866–1966.* London. HMSO.

Ministry of Defence 1973 *Report of the Defence Lands Committee 1971–73.* London. HMSO.

Moisley, H.A. 1962 The Highlands and islands — a crofting region? *Transactions of the Institute of British Geographers* **31**, 83–95.

Reconstruction Committee, Forestry Sub-Committee 1918 *Final report.* Cd. 8881. London. HMSO.

Royal Commission on Coastal Erosion and Afforestation 1909 *Second report.* Cd. 4460. London. HMSO.

Select Committee on Scottish Affairs 1972 *Land resource use in Scotland.* Report, Minutes of Evidence and Appendices, House of Commons Paper 511. London. HMSO.

Scottish Development Department 1986 *Land use change in Scotland 1985.* Statistical Bulletin No. 2(e) (1986), and annually. Edinburgh. Scottish Office.

Scottish Office 1992 *The Scottish Environment,* No. 3. Edinburgh. Scottish Office.

Scottish Office Agriculture and Fisheries Department (annually) *Agricultural statistics Scotland.* Edinburgh. HMSO.

Stamp, L.D. 1948 *The land of Britain: its use and misuse.* London. Longman.

Thomas, M.F. and Coppock, J.T. (eds) 1980 *Land assessment in Scotland.* Aberdeen University Press.

Tivy, J. 1973 Heather moorland. In J. Tivy (ed.), *The organic resources of Scotland,* 85–97 Edinburgh. Oliver and Boyd.

In: A. Fenton and D.A. Gillmor (eds) 1994 *Rural land use on the Atlantic periphery of Europe: Scotland and Ireland*, 55–73. Dublin. Royal Irish Academy.

IRISH LAND USE IN THE TWENTIETH CENTURY

Desmond A. Gillmor

Abstract: At the beginning of the twentieth century in Ireland a defective tenancy system had been replaced by almost universal owner-occupancy, but this system lacks flexibility and there is a small-farm problem. Apart from comprehensive agricultural statistics and a land use survey done in Northern Ireland in the 1930s, there is a dearth of information on Irish land use, though modern information technology is beginning to alleviate the situation. Four-fifths of Irish land is used by farmers and the agricultural usage is predominantly pastoral. Land-use trends in the twentieth century are shown and discussed, and comparisons between Northern Ireland and the Republic of Ireland are made. This is done with regard to tillage, cereals, milch cows, other cattle, sheep, horses and forestry. There has been increased intensification of land use, occurring earlier in Northern Ireland. Non-agricultural uses of land have increased in recent decades and this seems likely to continue.

Introduction

People's images of Ireland are composed essentially of rural landscapes, and in this respect perceptions of the country accord with reality. Although two-thirds of the population may now be classified as urban, outside of the Lagan corridor and the Dublin metropolitan area most people live in small towns, villages and the open countryside. To an even greater extent the land remains rural, comprising over 95% of the total area. The use to which this land is put is a vital element in the visual appearance of the Irish landscape, as well as providing the resource base for an important part of the economy. In setting the scene with regard to Irish land use in the twentieth century, this paper focuses on land ownership, information on land use, the present pattern of usage and, at greater length, land-use trends over the century.

Land ownership

Land utilisation is strongly influenced by the form of land ownership, since ownership largely confers the right of usage, so this aspect is considered briefly first. Control over land was a critical factor in Irish history, being the source of much of its conflict, and this still has a bearing on attitudes towards land. Patrick

Kavanagh's *Tarry Flynn* and John B. Keane's *The field* serve as illustrations from twentieth-century drama of the importance of land ownership in Irish rural society. There is a strong desire to own land, with emotional attachment to specific tracts of land. Such considerations of land ownership often overshadow the use to be made of the land, so that land may be neglected despite its perceived importance. There is the often-quoted comment by the New Zealand grassland expert Holmes (1949, 8) that he saw hundreds of fields growing just as little as it is physically possible to grow under an Irish sky. Nonetheless, features of Irish land use in the twentieth century, and especially during its second half, have been the increasing intensity of land use and growing competition for land from uses other than agricultural.

Land matters at the beginning of the twentieth century were dominated by the process of land reform, which constituted one of the most profound and influential movements in Irish history. A tenancy system with many deficiencies was at first improved and then abolished under a series of Land Acts from 1870. Transfer of ownership from the landlords to the occupying tenants had been largely effected by the time of partition, but was completed compulsorily under legislation in the Irish Free State in 1923 and in Northern Ireland in 1925. The resultant universal system of owner-occupancy ensures that land use is under the control and personal interest of the individual owners. Transfer is predominantly by inheritance, with only about one-fifth of it through the open market, so that many farms remain in the same family for generations. The immobility in land ownership, combined with the cost of land, limits the opportunity to acquire land for young people without farms and for smallholders wishing to expand.

Less than one-tenth of Irish land is tenanted, the lowest proportion in the European Community (EC). It is predominantly let for periods of less than one year under the conacre system by people who do not wish or are unable to work it themselves, the practice being most common in the north (Mac Aodha 1967). Conacre provides a market in land use, but the short-term nature of the letting is a disincentive to land improvement. Longer-term leasing had been discouraged as a reaction against the defects of the former tenancy system but it is now promoted, with very limited effect, as a means of access to land use for those who lack capital or opportunity for purchases.

The immobility of the particular owner-occupancy system of land tenure which evolved has been a major factor, with others, contributing to the relatively slow trend towards farm enlargement. Despite the disappearance of many of the smallest farms, the mean area of crops and pasture per holding over 0.4ha increased only from 12.0ha in 1900 to 17.6ha in 1980. Land usage varies to some extent with farm size — for instance, wheat-growing and fattening of beef cattle are associated mainly with large holdings. However, the farmyard pig and poultry production which was suited to the needs of the small farm has been lost to large-scale production units under capital-intensive management systems. An overall trend in recent decades has been for a disproportionately greater share of agricultural output to be contributed by the larger farms, especially with regard to the more profitable enterprises of dairying and tillage. This accentuates the disadvantageous situation of the many small farms which often have inappropriate land-use systems and are unable to afford an adequate income for those owners fully dependent upon them. The small-farm problem is a crucial one in Irish agriculture and has a distinct spatial dimension, farm size diminishing westwards and northwestwards. The problem is accentuated by the tendency for the smaller farms to be on the poorer land, so that their land resources are limited both in quantity and in quality.

Of the land area of Ireland, about 80% is contained within agricultural holdings. The family farm is the basic unit of land ownership. Only 0.5% of landholders were commercial or institutional in the Republic of Ireland in 1980. These were mainly organisations or institutions such as sports clubs, schools, colleges, hospitals, government bodies and cooperatives, but also property development companies. They were mainly in the vicinities of the larger towns, with the main concentration being in County Dublin (Horner *et al.* 1984). Corporate involvement in Irish agriculture seems to be discouraged by low returns on capital investment, fluctuations in profitability, the complexities of management and small farm size.

Data on the ownership of land not on agricultural holdings are grossly deficient. A substantial area is associated with farm holdings in that it is under the joint ownership of farmers in the locality who have the right to graze certain numbers of livestock on it. Such commonage is usually mountain, bog or other rough grazing, and it probably accounts for nearly one-tenth of the land. The largest landowners in Ireland are the state forest authorities, which together own 470,000ha or 5.7% of the total land area. Another major landowner in the Republic of Ireland is the semi-state peat authority, Bord na Móna, which owns 88,000ha or 0.12% of the land. The ownership of land, whether farmland or non-agricultural land, is predominantly Irish, with only a small proportion being under foreign control.

Information on land use

The information needed to investigate present Irish land use and that throughout the twentieth century is quite variable. Availability is confined largely to the statistics of the agricultural enumeration or census. These exclude most other uses but afford detailed information on crops and livestock for much of the land area of the island and are very unusual in providing an almost continuous series of data since 1847, twenty years earlier than in Britain. Complete returns had been made by the police up to 1918 but subsequent conditions interrupted this, so that for the period from 1919 to 1922 in Northern Ireland and to 1924 in the Irish Free State estimates were based on sample returns from farmers and the data are less reliable or comprehensive.

In the Republic of Ireland during the period 1960–80 complete agricultural enumerations were done at five-year intervals; the results were published by county and rural district, but access to unpublished district electoral division figures was granted by the Central Statistics Office. In intervening years there was a partial enumeration in June using sampling procedures, and statistics were published by county; crop data were not available for the years 1977–9. Since 1980 the only complete enumeration was in 1991, and for other years statistics have not been published at even the county level. They are available only for five regions which incorporate three to seven counties and have considerable internal heterogeneity. Thus there has been a significant deterioration in the availability of statistical data on agricultural land use in the Republic of Ireland.

In Northern Ireland there is an annual agricultural census with statistics available on county and rural district bases. Since 1981, however, only holdings above a minimum size specified in terms of area, business and labour have been included, numbering about 41,000 at present. Estimates for restricted features based on a three-year rotational survey are made for the other 14,800 'minor' holdings, which occupy less than 4% of the agricultural land. Such alterations in census practice and also definitional changes and discordance in statistics hinder investigation of trends in land usage over time and affect comparability of data between Northern Ireland and the Irish Republic.

The availability of the agricultural statistics has enabled the cartographic representation of spatial patterns of Irish land use, showing the extent to which variation can occur even within a small country characterised by mixed livestock farming. The first major collection of maps related to the late 1920s and was contained in *An agricultural atlas of Ireland* (Stamp 1931), an early classic work of this type. The core of the atlas comprised dot distribution maps and associated commentaries, eight for Northern Ireland and eleven for the Irish Free State. Agricultural distributions in the Republic of Ireland in 1970 were mapped by Gillmor (1977). The most comprehensive cartographic representation of land use was by Horner *et al.* (1984), relating to all of Ireland in 1980 and including also many maps showing change over the decade 1970–80. The maps in each of these publications were compiled from the statistics for rural districts, but those for district electoral divisions, the lowest level of aggregation, have been mapped for some counties and regions. These works include those by Johnson and MacAodha (1967) and Ó Cinnéide and Cawley (1983), and they show the variation in land use that can occur even within rural districts. General books focusing mainly on rural Ireland and containing material on the utilisation of land include those edited by Gillmor (1979; 1989), Cruickshank and Wilcock (1982), Jess *et al.* (1984), Aalen (1985), Breathnach and Cawley (1986), Montgomery *et al.* (1988) and Feehan (1992).

Mapping of actual land use, rather than that aggregated by artificial statistical areas, has until recently been almost absent in Ireland, with one very notable exception — the Land Use Survey of Northern Ireland. This was initiated in 1936 when the Geographical Association of Northern Ireland undertook field mapping of the land use of the Belfast area, having been stimulated by the work being done by Stamp in the first land utilisation survey in Britain. The work was extended to County Antrim and then to all of Northern Ireland, through the efforts of about a thousand voluntary field-workers. Mapping was mainly undertaken in the years 1937–9 portraying a picture of Northern Irish land use prior to the Second World War. In the period 1945–51 the field record on six inches to one mile (1:10,560) maps was reduced to one inch to one mile (1:63,360) and published in colour by the government. It had been planned to produce explanatory memoirs for each of the sheets but this was done only for the Belfast area (Hill 1948). Instead, a book summarising and analysing the pre-war land-use pattern for all of Northern Ireland and tracing subsequent trends was edited by Symons (1963).

Such a record of land use does not exist for the Republic of Ireland. In the early 1950s a proposal was made for a land-use survey of the state using volunteer members of the then rural organisations Muintir na Tire and Macra na Feirme (Lane 1953). A pilot survey of parts of Counties Kildare and Tipperary was initiated but the survey did not proceed. A second proposal was made by the Association of Geography Teachers of Ireland in the 1960s, at the time of the second land-utilisation survey in Britain. The Association saw the educational value of involving school students in the mapping of land use. Approaches were made to government but the necessary finance was not forthcoming.

Such ground mapping of land use involves huge labour input extending over a considerable time, but the advent of aerial photography and satellite imagery dispensed with the need for much of this, though some field checking under varied land-use conditions is an essential part of any programme. The storage of remote sensed material in digital form through the application of geographical information systems greatly facilitates the processing of large volumes of land-use data. While a land-cover map of Ireland on one sheet was produced from LANDSAT imagery, the modern technology has been applied mainly in localised studies of a specific nature. Examples include the study of land-cover change in Areas of Outstanding Natural Beauty in Northern Ireland, the mapping of land use in an investigation of water acidification in Connemara, and the monitoring of peatland in Northern Ireland and of arable land in County Meath. Some of the potential of the modern technology in the Irish context was demonstrated in a supplement to *Technology Ireland* (Kennedy and Mac Siúrtain 1987) and in the proceedings of the Geographical Society of Ireland annual conference in Coleraine in 1991 (Cooper and Wilson 1992). In general, remote sensing is much more applicable to land form or cover, and hence to ecological studies, than to land activity or function as required in land-use studies. Obviously land cover and usage are often interrelated, though a particular cover might have one of several different uses or a combination together in multiple land use.

The major undertaking in Irish land studies, which encompasses all of the island, is the CORINE Land Project (Ireland) (O'Sullivan 1992). The CORINE Programme was established by the EC in 1985 to provide an environmental information system for the Community. One of its data bases is the Land Cover Project, which seeks to provide quantitative data on land cover which are consistent and comparable across the Community and to provide one land-cover data base at a scale of 1:100,000 using the 44 class CORINE nomenclature. The application of a standard methodology involves computer-assisted photointerpretation and the simultaneous consultation of ancillary data. The CORINE Land Project (Ireland) was initiated in March 1992 and the cartographic product will be a land-cover map of the island on the 1:100,000 scale. The project is led by the Ordnance Survey of Ireland, coordinating with the Ordnance Survey of Northern Ireland, and it involves universities and government departments as partners in the project. Satellite imagery from 1989–90 is classified and digitised, with the data being held in a GIS Arc/Info system. The CORINE nomenclature seems in some respects to be much better suited to Mediterranean than to Irish or Scottish circumstances, but adaptations in relation to pastures and peat bogs have been made in the Irish context. The restriction to a minimum 25ha area to be mapped limits usefulness at the local level in the context of the fragmented nature of Irish land use, and there are always questions relating to the accuracy of satellite imagery and its interpretation. While concerned specifically with land cover, the CORINE project is a major source of information relevant to Irish land use, all the more so because of the lack of alternative material. This dearth is to be deplored from an academic perspective but is much more critical for policy-makers.

Land-use pattern

Portrayal of the present land-use pattern must still be based on the statistics of the agricultural census and these incorporate in part a land-cover element. The structure of land use in Ireland as a whole in 1990 is shown in Fig. 1, where it may be compared with that in 1900, though trends are considered in the next section. Land use in Northern Ireland and the Republic of Ireland in 1990 is given, in so far as the statistics permit, in Table 1. The data for Northern Ireland had to be derived through calculations and estimations because of the nature of the agricultural census, including the exclusion of 'minor holdings' from the main census and the absence of a distinction between grassland for mowing and for grazing since 1987.

The extent to which Irish land use is dominated by agriculture is evident, four-fifths of the land being used by farmers. This comprises both the area of crops and pasture, or improved land, and the unimproved land used as rough grazing. The distinctions between these two categories and between rough grazing and some 'other land' are often not clear, as one may grade into another, and what constitutes 'pasture', 'rough grazing' and 'other land' depends to some extent upon the interpretations of the individual farmers and census enumerators. The rough grazing is used principally for sheep and some cattle, mainly in the uplands. The improved land occupies two-thirds of Ireland but its distribution varies spatially, comprising almost the entire countryside in some areas (Fig. 2). The gaps in the pattern of improved land correspond mainly with the uplands and peat bogs and are most pronounced in the west.

The fundamental distinction in the use of the improved agricultural land is that between tilling it for arable crops and having it under grass. Its predominantly pastoral nature is the most distinctive feature of Irish land use. Grassland accounts for 92% of the area of crops and pasture and, combined with rough grazing, it occupies 76% of the total land area of the island. It is grazed mainly by beef cattle,

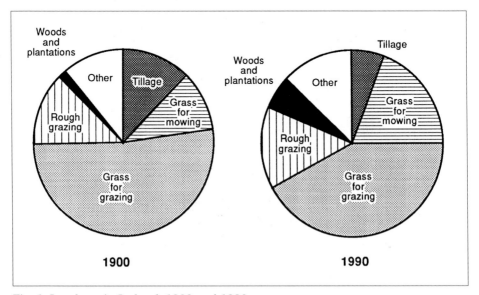

Fig. 1. Land use in Ireland, 1900 and 1990.

TABLE 1. Land use in Ireland, 1990.

Land uses	Republic of Ireland		Northern Ireland		Ireland	
	ha ('000)	%	ha ('000)	%	ha ('000)	%
Crops and pasture	4682.5	68.0	864.0	63.7	5546.5	67.3
tillage	415.9	6.0	66.7	4.9	482.6	5.9
grass for mowing	1287.7	18.7	286.2	21.1	1573.9	19.1
grass for grazing	2978.9	43.2	511.1	37.7	3490.0	42.3
Rough grazing	966.1	14.0	198.8	14.7	1164.9	14.1
Woods and plantations	407.0	5.9	73.0	5.4	480.0	5.8
Other land	833.6	12.1	220.6	16.3	1054.2	12.8

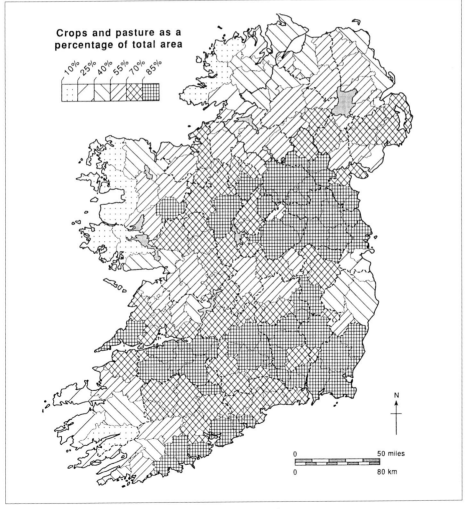

Fig. 2. Distribution of improved agricultural land.

dairy cattle and sheep, with some horses and minor livestock. Grass is also the major source of winter feed, 31% of the grassland being conserved as silage or hay. The strong emphasis on grass in Irish agriculture is related to the moist and mild climate, favourable edaphic conditions, a strong pastoral tradition in Irish rural society, and the market outlets for livestock and their products.

Less than one-tenth of the improved land is tilled, but there are striking spatial variations in the extent of arable cropping (Fig. 3). Tillage tends to be most common where the climate is driest and sunniest, where soils are well drained, where farms are larger and where the external historical influences of landlords and tenant farmers were greatest. The tilled land is devoted primarily to cereals, which account for 78% of it, with root and green crops occupying 21% and horticulture 1%. Cereal crops are dominated in turn by barley, 73% of the total, followed by wheat and oats. The principal root crops are potatoes and sugar-beet.

The non-agricultural land includes forest, mountain heath, peat bog, bare rock, quarries, water and recreational land, but also buildings, roads and urban

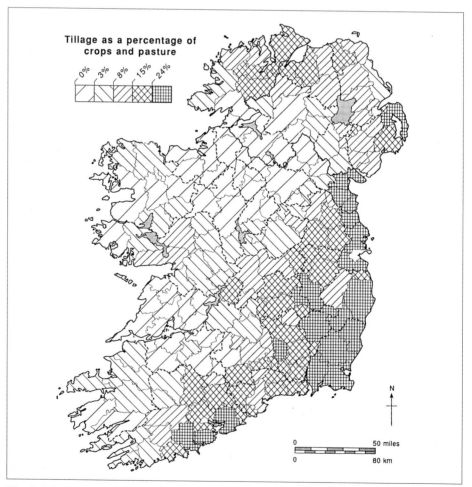

Fig. 3. Distribution of arable cropping.

areas. In contrast with the agricultural land, statistical data on the specific uses of this land are almost non-existent, apart from forestry. Woods and plantations occupy 6% of the land cover, so that Ireland is the least forested country in Europe apart from Iceland. Much of the land owned by Bord na Móna is used for mechanised peat production, mainly on the raised bogs of the midlands and in parts of the west, and in addition there is private cutting of peat on an unquantifiable area of bog. Quarrying is a conspicuous but localised land use, especially in areas of glacial sand and gravel near to urban centres of demand by the construction industry. Water catchment and reservoirs constitute an important land use in upland districts near to urban areas, principally to supply Dublin from the Wicklow and Dublin Mountains and Belfast from the Mourne Mountains and Antrim Plateau. Reservoirs also serve hydroelectric purposes on the Rivers Liffey, Lee and Erne. Use of land specifically for recreation only is a tiny proportion of the total and is most obvious in relation to golf courses and sports pitches. Much more extensive is the recreational component in multiple land use, as in forest and farm recreation, hill-walking, hunting, game-shooting and angling. Such multiple usage, as by recreation with forestry, agriculture, water catchment, conservation and other uses, would make statistical specification difficult even if basic data were available. Alexander and Gahan (1987) estimated that only 0.5% of the land area of the Republic of Ireland was protected for nature conservation, over two-thirds of this in the three National Parks at that time and the remainder in small nature reserves. In Northern Ireland the percentage of land occupied by buildings and roads ranged from 4.3 in Fermanagh to 6.8 in Down in the 1950s (Symons 1963). Omitting rural roads and buildings and reflecting in part its lower level of urbanisation, the Irish Republic's statutorily defined urban area was only 0.9% of the state in the 1970s, but this underestimated by one-half the built-up area in towns for which more precise data could be obtained (Bannon 1979).

The pattern of rural land use is a major component in the appearance and character of the Irish landscape. The proverbial greenness of the Irish countryside noted by many visitors is attributable mainly to the extent of grassland and its luxuriance. Much of the human imprint on the land has been that associated with farming, and spatial variations in agriculture are a major element in the diversity of regional landscapes. Examples of landscapes of fundamentally different character related to differences in land-use patterns include the pastoral dairying landscape of County Limerick, the mixed arable and grassland landscape of County Carlow, and the landscape of the western seaboard with its intermixture of improved land, rough grazing and non-agricultural land.

Trends in land use

Land use is a dynamic phenomenon, and the present pattern of Irish land use is the outcome of a long evolution in the interaction between humans and their environment. The major trend prior to the twentieth century and dating from Neolithic times was the spread of farmland at the expense of forest. The native forest cover was cleared not only for agriculture and settlement but also for timber production, and it had been affected by the spread of peat bogs, so that by 1900 only about 1.5% of the land was forested. The area of improved agricultural land reached its peak not under the population pressure of pre-Famine times but in 1875, before the decline initiated by the agricultural recession consequent upon the import to Europe of New World produce. It then accounted for 77.4% of the total land area. Within the improved agricultural area, a major shift from arable to pastoral usage was occurring. Between 1851 and 1900 the area of tillage diminished by 46%, from 34.7% to 16.4% of the improved land. Cereals were most affected, especially wheat. Cereal prices had fallen after repeal of the Corn Laws in 1846 and with growing imports to Britain from foreign sources. The prices for livestock increased as demand for meat in Britain expanded and transport to this market was improved. These price trends encouraged an increasing emphasis on pastoral farming, reinforced by the declining farm labour force and home consumers of arable products as the Irish population fell. Although the cow herd was static, the number of other cattle increased substantially, with total cattle first exceeding the human population of Ireland in the early 1890s. Changes in agricultural land use in the twentieth century have to some extent been a continuation, but at a diminished rate initially, of the trends established in the preceding century. The period until the 1960s was seen by Crotty (1966) as one of stagnation in the agriculture of the Irish Republic, resulting from the establishment of owner-occupancy.

The twentieth-century trends in the major features of agricultural land use are shown in Figs 4–8. These graphs, apart from Fig. 5, show values as percentages of those in the base year 1900, in part so that the magnitudes of change in the different aspects of land use can be compared more readily but also to facilitate comparison between the Republic of Ireland and Northern Ireland. This comparison is of particular interest because the two territories were within the same political unit of the United Kingdom (UK) until the partition of Ireland in 1922 resulted in different policy contexts, but since accession to the EC in 1973 both have come within the same policy regime of its Common Agricultural Policy (CAP). The divergence and convergence in agricultural development paths do not correspond precisely with these dates, however, as divergence was greatest in the 1930s when the Irish Free State adopted a policy of self-sufficiency and there was an economic war with Britain, and convergence had begun with agricultural modernisation in the Republic of Ireland from about 1960.

In the graphs the livestock are measured in terms of livestock unit equivalents, which take into account the varying feeding capacities of the different ages and sizes of animal, and are thus a more meaningful measure of land-use impact than are total numbers. The trends in milch cows and other cattle may be taken as representative of dairying and beef production respectively, in the absence of more precise indicators, though the milch cows include beef cows and some of the other cattle are dairy herd replacements. The livestock trends constitute major elements in land-use change because of the predominantly pastoral nature of Irish land. A perspective different from the trend graphs but reflecting changes in land usage is given by a comparison of the structure of agricultural output at different times (Table 2).

The most important trend in Irish agricultural land use in the twentieth century has been a continuation from 1851 of the tendency for the area of arable crops or tillage to decline under normal circumstances (Fig. 4). As in the nineteenth century, this has been related to market and price trends and declining population; with grain available from places where its comparative

TABLE 2. Percentage structure of the value of gross agricultural output.

Farm enterprises	Republic of Ireland			Northern Ireland		
	1926	1960	1990	1926	1960	1990
Tillage	14.6	20.4	12.4	21.6	9.1	6.2
Milk	24.1	23.1	33.4	19.4	17.0	27.4
Cattle	24.3	30.3	39.7	23.2	21.1	36.1
Sheep	5.0	7.0	4.9	4.3	5.8	9.4
Pigs	16.0	11.4	5.7	9.3	28.3	10.1
Poultry	15.9	7.7	4.0	22.3	18.7	10.8

advantage was greater, Irish farmers have tended to revert to the environmentally more favoured livestock production. The downward trends were reasserted quickly after the two abrupt and major interruptions resulting from compulsory tillage during the two world wars but with a time-lag effect. Political circumstances account also for the marked divergence between the two territories which was maintained from its initiation in the early 1930s. This was when encouragement was given to tillage by the government of the Irish Free State as part of its self-sufficiency strategy and its effort to maximise rural employment, while the

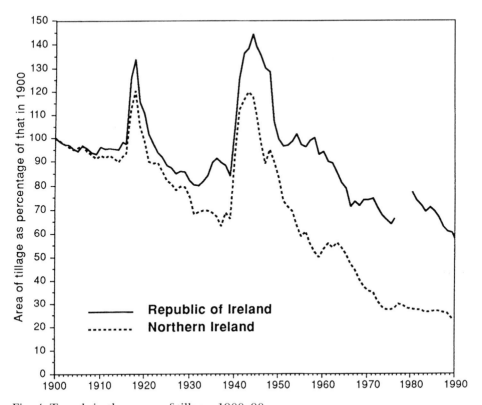

Fig. 4. Trends in the areas of tillage, 1900–90.

cattle industry suffered difficulties because of the economic dispute with Britain. Protectionism resulted in high prices for cereals while Northern Irish farmers had access to feedstuffs at low world prices. This was a major factor contributing to huge growth in the pig and poultry industries there as compared with the Irish Republic, leading to perhaps the most distinctive difference in the agriculture of the two territories. The areas of tillage in both stabilised under the protected market and price regime of the EC at first, but fell more recently in the Irish Republic in particular.

Within the diminishing tillage area, substantial changes occurred in the relative importance of individual crops. Cereals have assumed an increasingly dominant role, rising from 54% of the tilled area in 1900 to 78% in 1990. The area of root and green crops has diminished much more rapidly, mainly because of the large labour requirement in their production and in feeding to animals, and the declining demand as they were replaced by compounded livestock feedstuffs and grass and as human consumption of potatoes fell. The 36,500ha of potatoes grown in 1990 was only 14% of the area in 1900. The decline in the area of root and green crops occurred despite the introduction of sugar-beet to supply a protected sugar industry in the Irish Free State in the 1920s and 1930s, with 32,300ha being grown there in 1990 and none in Northern Ireland. Flax was a crop which formerly had been important in Ulster but, having reached wartime peaks, it almost disappeared in the 1950s. Crop rotations have become much simpler, with much more successive cropping of cereals.

Strikingly different trends have occurred in the areas of the major cereals, as illustrated by those in the Republic of Ireland (Fig. 5). Most prominent has been the substitution of barley for oats over the last half-century. Oats had been the dominant cereal, occupying 55% of the area as late as 1950, whereas barley accounted for 72% of cereals in 1990. The decline of oats was related in part to the decrease in the number of horses and farmyard poultry, for both of which they were an important feedstuff, and to the reduced role of oat products in the human diet. More important, however, was the increased comparative advantage of feeding barley, with the introduction of higher-yielding varieties, development of short-strawed strains suitable for combining, improved liming and fertilising of soils, acceptance by farmers of the superior yield and feeding properties of barley, growth of feedstuff compounding, and establishment of minimum price levels. Almost all wheat requirements had been imported but, following the adoption of protection and subsidy, its price was twice the world level by 1936. The area of wheat continued to rise steeply in wartime conditions to a peak of 268,100ha in 1945, the largest area since agricultural statistics were first recorded. There has been subsequent decline, with considerable fluctuations, but wheat remains much more important than in Northern Ireland.

The striking feature of the trends in milch cows is the dramatic growth in numbers within the short period from the late 1950s until 1974, preceded by overall lack of expansion and succeeded by some decline (Fig. 6). At times during the first half of the century, however, trends were not uniform throughout the island. The early decades were an unfavourable time for the Irish dairy industry, but the major development of the creamery industry in Munster provided a market for milk which enabled farmers to expand their herds (O'Donovan 1940). Some of this comparative gain to the area which became the Irish Free State was lost during the difficulties there in the 1930s. It was the

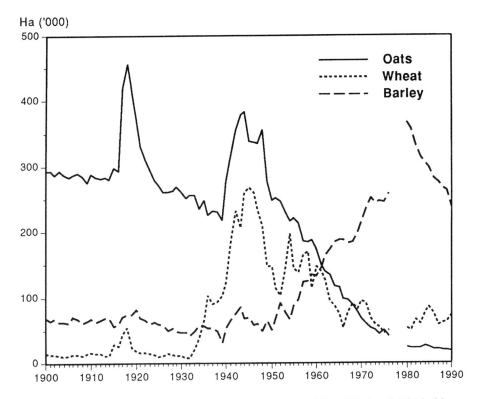

Ha ('000)

Fig. 5. Trends in the areas of cereal crops in the Republic of Ireland, 1900–90.

period of the Second World War, however, in which comparative growth was greatest in Northern Ireland. Cow numbers increased there by over one-quarter as a result of the wartime priority and financial incentives given to milk production in the UK, but with over-production the gain was lost rapidly in the early 1950s. The later rapid expansion was in response to increasing prices for milk, boosted by the huge financial support given to the dairy industry by both governments, and growing demand for cattle. Farmers were attracted by the regular, stable and high incomes from dairying, which was actively promoted by government, advisory service and mass media and benefited from research developments. Growth was terminated by a crisis in the cattle industry in 1974, by agricultural recession in the late 1970s, and by the imposition of the milk superlevy in 1984.

Though somewhat moderated, the trends in other cattle bear resemblances to those for milch cows because of the close interlinkages between the dairy and beef industries, the former supplying many of the replacement stock to the latter (Fig. 7). Thus the trends in the early part of the century reflected to some extent those in milch cow numbers, and also both were affected by high mortality because of very wet weather in 1924. There had been difficult marketing conditions in the 1920s, but the cattle industry of the Irish Free State reached its lowest ebb in the 1930s, when it was the agricultural sector most adversely affected by the economic dispute with Britain and policy discrimination against

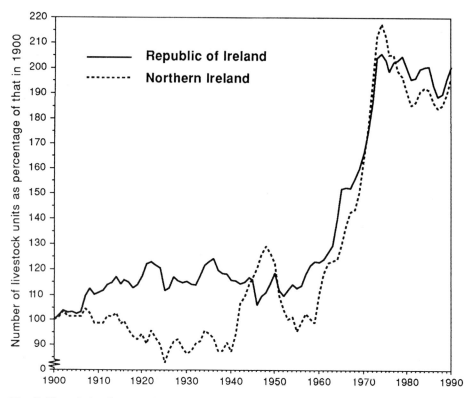

Fig. 6. Trends in the numbers of milch cow units, 1900–90.

livestock. Trends diverged at this time, with decline in the Irish Free State and growth in Northern Ireland, where greater growth in the 1940s was made possible by the wartime expansion in the breeding herd. Beginning in the 1930s, state support in Northern Ireland came much earlier and at a higher level than in the Irish Republic, but southern farmers later benefited through sale of store cattle to the north, boosting numbers there. Growth in the number of Irish cattle prior to 1960, while the breeding herd was generally static, was made possible by reduction in the mortality of calves. Northern Ireland maintained much of its lead through the growth period, and reversal after the cattle crisis of the mid-1970s was less than in the Republic of Ireland.

The trends in sheep numbers in the two parts of Ireland match each other in many respects, but it is possible to recognise a tendency, apparent at times in the other land-use trends also, for Northern Ireland farmers to respond to stimuli more quickly and to a greater extent than their southern counterparts (Fig. 8). It is not easy to explain fully the divergence in trends, though influential factors were the much earlier and greater state support of the sheep industry in Northern Ireland and the associated intensification in the use of hill land there. Although divergence had begun earlier, conditions in the 1930s played a major role. Sheep numbers declined in the Irish Free State during the difficulties in its livestock sector at that time, while they increased substantially in Northern Ireland. Subsequent trends reflected competition from other enterprises on

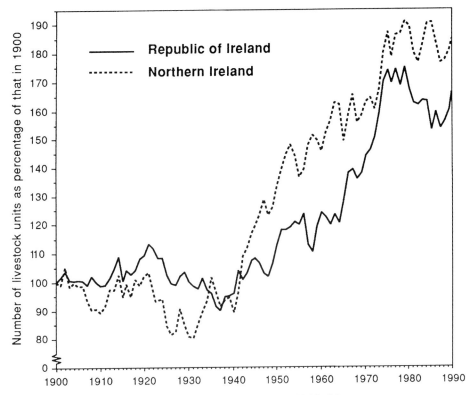

Fig. 7. Trends in the number of other cattle units, 1900–90.

lowland farms where sheep tended to be a subsidiary activity, reflecting the interrelationships between land uses. Thus numbers diminished during the wartime conditions, when sheep gave way to tillage. Decline was greater in Northern Ireland, where there was additional competition for land from the expanding dairy herd. Mountain flocks in both territories suffered under exceptionally severe weather conditions in the winter of 1946–7. Numbers then recovered and rose under the influence of favourable trading conditions and government grants. The decline which followed from the early 1960s was mainly on the lowlands, where sheep gave way to the expanding dairy and beef herds. The industry suffered from comparative lack of state support and technical advance relative to other enterprises. Circumstances changed with the adoption of an EC common sheep policy in 1980 and later curtailment of milk production. The resultant huge growth in sheep numbers constituted one of the most dramatic changes in the history of Irish land use.

A trend which resulted from the changing technology of agricultural production rather than being a part of the composition of output was the decline in the number of working horses. This began earlier in Northern Ireland than in the Irish Free State, where in 1945 there were still 445,500 horses and ponies, accounting for 15% of grazing livestock units. In 1990 there were 53,500 horses and ponies, almost all associated with the breeding industry and recreational purposes. The decrease in the number of horses had substantial implications for

Ha ('000)

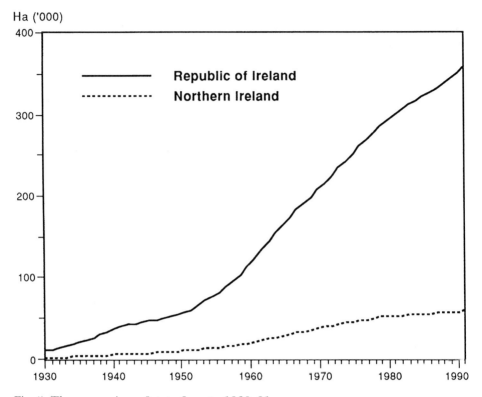

Fig. 9. The expansion of state forests, 1930–91.

catchment increased with the demands for their products, especially from urban areas. Growing recreational pressure has been mainly on land used also for other purposes rather than involving wholesale land use conversion, though there has been establishment of golf courses, playing fields, car-parking facilities, and camping and caravan sites. Conservation has become a land use only in recent decades. The area under rural buildings has increased, both in modern farm structures and in residential development, the extent of the latter in the open countryside being particularly great under the less stringent physical planning of the Republic of Ireland. Land use conversion for forestry, peat production and conservation has been greatest in the areas of greatest rurality and remoteness, but the high environmental quality of these places renders such change all the more significant. It is in the urban fringe areas, however, that the pressures of competing uses such as building, transport, recreation, quarrying and waste disposal have become most strongly felt.

Conclusion

There are deficiencies in the information on rural land use in Ireland, but the data available have been used to outline the present pattern and trends in the twentieth century. This indicates that shifts have occurred between land uses but nonetheless that there has been considerable stability in the overall pattern, which is dominated by agriculture. There has been substantial growth in other land uses in recent decades and this seems likely to increase further in the remainder of the century within the context of changing EC policy.

References

Aalen, F.H.A. (ed.) 1985 *The future of the Irish rural landscape*. Department of Geography, Trinity College, Dublin.

Alexander, R.W. and Gahan, S.V.E. 1987 A review of sites protected for conservation in the Republic of Ireland. *Irish Geography* **20** (2), 82–8.

Bannon, M.J. 1979 Urban land. In D.A. Gillmor (ed.), *Irish resources and land use*, 250–69. Dublin. Institute of Public Administration.

Breathneach, P. and Cawley, M. (eds) 1986 *Change and development in rural Ireland*. Geographical Society of Ireland, Special Publications No. 1. Maynooth.

Cooper, A. and Wilson, P. (eds) 1992 *Managing land use change*. Geographical Society of Ireland, Special Publications No. 7. Coleraine.

Crotty, R.D. 1966 *Irish agricultural production: its volume and structure*. Cork University Press.

Cruickshank, J.G. and Wilcock, D.N. (eds) 1982 *Northern Ireland environment and natural resources*. The Queen's University of Belfast and the New University of Ulster.

Feehan, J. (ed.) 1992 *Environment and development in Ireland*. Environmental Institute, University College Dublin.

Gillmor, D.A. 1977 *Agriculture in the Republic of Ireland*. Budapest. Akadémiai Kiadó.

Gillmor, D.A. (ed.) 1979 *Irish resources and land use*. Dublin. Institute of Public Administration.

Gillmor, D.A. (ed.) 1989 *The Irish countryside: landscape, wildlife, history, people*. Dublin. Wolfhound.

Hill, D.A. 1948 *The land of Ulster I. The Belfast region*. Belfast. HMSO.

Holmes, G.A. 1949 *Report on the present state and methods for improvement of Irish land*. Dublin. Stationery Office.

Horner, A.A., Walsh, J.A. and Williams, J.A. 1984 *Agriculture in Ireland: a census atlas*. Department of Geography, University College Dublin.

Jess, P.M., Greer, J.V., Buchanan, R.H. and Armstrong, W.J. (eds) 1984 *Planning and development in rural areas*. Institute of Irish Studies, Queen's University, Belfast.

Johnson, J.H. and MacAodha, B.S. 1967 *An agricultural atlas of County Galway*. Social Science Research Centre, University College Galway.

Kennedy, T. and Mac Siúrtain, M. 1987 Remote sensing and image analysis at University College, Dublin. Supplement to *Technology Ireland* (February 1987).

Lane, E.V. 1953 Land use surveys. *Rural Ireland* (1953), 3–11.

MacAodha, B.S. 1967 *Conacre in Ireland*. Social Sciences Research Centre, University College Galway.

Montgomery, W.I., McAdam, J.H. and Smith, B.J. 1988. *The high country: land use and land use change in Northern Irish uplands*. Belfast. Institute of Biology.

Ó Cinnéide, M.S. and Cawley, M.E. 1983 *Development of agriculture in the west of Ireland*. Blackrock. An Chomchairle Oilúna Talmhaíochta.

O'Donovan, J. 1940 *The economic history of livestock in Ireland*. Cork University Press.

O'Sullivan, G. 1992 A new view of Ireland. *Technology Ireland* (July/August), 32–5.

Sheehy, S.J., O'Brien, J. T. and McClelland, S.D. 1981 *Agriculture in Northern Ireland and the Republic of Ireland*. Dublin and Belfast. Co-operation North.

Stamp, L.D. 1931 *An agricultural atlas of Ireland*. London. Gill.

Symons, L. 1963 *Land use in Northern Ireland*. University of London Press.

Symons, L. 1970 Rural land utilisation in Ireland. In N. Stephens and R.E. Glasscock (eds), *Irish geographical studies in honour of E. Estyn Evans*, 259–73. Department of Geography, Queen's University, Belfast.

In: A. Fenton and D.A. Gillmor (eds) 1994 *Rural land use on the Atlantic periphery of Europe: Scotland and Ireland*, 75–98. Dublin. Royal Irish Academy.

AGRICULTURE AS A LAND USE IN SCOTLAND

Guy M. Robinson

Abstract: This paper outlines briefly some of the principal factors affecting the character of agriculture in Scotland. It then examines the distribution of the main agricultural land uses and concludes with a consideration of changes that have occurred over the last twenty years under the influence of the Common Agricultural Policy (CAP) of the European Community (EC). Computer maps employing data from the annual Agricultural Census are used to show the distributions of some of the principal crops and livestock on Scottish farms. These distributions reflect the strong physical controls upon Scottish agriculture, with clear divisions between the lowlands, the Southern Uplands, and the Highlands and Islands. The growing importance of cereal production and oilseed rape under the CAP price support system is emphasised. The paper concludes with a review of recent measures adopted as reforms to the CAP.

Introduction

In 1991 the area recorded in the Agricultural Census as being under crops, fallow, grass and rough grazing in Scotland was 5.7 million ha, or 73.8% of the land surface. Yet, despite its continuing dominance of Scottish land use, agriculture's contribution to the national gross domestic product (GDP) is small. Its share is also declining, from just over 4% of GDP in 1972 to less than 2.5% in 1990. Agriculture's share of the national labour force is also small — around 1.5% of those in employment. Nevertheless, in 1990 agriculture accounted for £1569.2 million of output and contributed to both output and employment in related industries, especially the food-processing and farm supply sectors.

As shown in Table 1, the biggest returns in Scottish farming are from fat cattle and calves, followed by milk and milk products, fat sheep and lambs, barley, potatoes, wheat, poultry, fat pigs and oilseed rape. Livestock and livestock products account for just over two-thirds of farm output, but this proportion has fallen since the UK entered the EC in 1973. Comparisons between recent data and those for 1972–3 are apposite, representing changes occurring under the influence of the CAP, and so several such comparisons are made in this paper. In terms of output, the largest change in the last two decades has been the increase in returns from sales of crops. Favourable price supports, especially for cereals and oilseed rape, have encouraged this sector of farming whilst, in contrast,

75

TABLE 1. Farm output and expenditure, 1972 and 1990.

	1972 Output		1990 Output	
	£ mill.	%	£ mill.	%
(a) *Livestock*				
Fat cattle & calves	71.5	25.3	376.5	24.0
Fat sheep & lambs	22.6	8.0	197.8	12.6
Poultry	12.4	4.4	81.6	5.2
Fat pigs	19.8	7.0	58.3	3.7
Others*	20.9	7.3	65.7	4.2
All livestock	147.2	52.1	779.9	49.7
(b) *Livestock products*				
Milk & milk products	57.8	20.4	245.9	15.7
Eggs	14.6	5.5	33.8	2.2
Clip wool	3.8	1.2	10.8	0.7
Others	1.1	0.3	6.9	0.4
All livestock products	77.3	27.4	297.4	19.0
(c) *Crops*				
Barley	21.4	7.6	162.9	10.4
Potatoes	14.0	5.0	99.6	6.3
Wheat	4.3	1.5	85.4	5.4
Oats	3.8	1.3	10.9	0.7
Oilseed rape			44.8	2.9
Others	1.1	0.4	11.7	0.8
All crops	44.6	15.8	415.3	26.5
(d) *Horticulture*	11.4	4.0	60.5	3.9
Miscellaneous	2.1	0.7	16.1	1.0
Total output	282.6	100.0	1569.2	100.0

* Consists largely of sales of store cattle and store sheep.

Sources: *Scottish Agricultural Economics* 23 (1973); DAFS 1973; *Scottish Abstract of Statistics* 9 (1980); *Scottish Abstract of Statistics* 19 (1990); SOAFD 1991.

quotas have restricted milk production. Milk's share of farm output has fallen as a result, whilst the shares of output supplied by wool, eggs, pigs and the traditional Scottish cereal, oats, have also declined.

Factors affecting agricultural production

The distribution of agricultural land closely reflects the strong climatic constraints on Scottish agriculture, which in turn reflect the control of relief:

14% of the country lies above 450m and two-thirds is above 300m (Fig. 1a). Most of the land above 150m is rough grazing and woodland, with the upper limit of cultivation being reached in many parts of the Highlands. This limit is much lower in the west; for example, in the West Highlands there is little cultivated land above 30m. Large areas of western Scotland receive over 1500mm of rainfall per annum, and have fewer than 1000 day-degrees centigrade (Birse and Dry 1969; Miller 1973). This harshness severely restricts the range of possible agricultural enterprises. Physical constraints are fewer, however, in the east, the Central Lowlands and some of the coastal lowlands of the west, especially the southwest. The extent of physical restrictions on Scottish agriculture is indicated by the substantial area of the country designated under EC guidelines as a Less Favoured Area (LFA), in which farmers can receive compensatory payments for farming in an area of 'environmental difficulty' (Fig. 1b).

The strong influence of relief and climate upon the distribution of the best soils for agriculture is indicated in Fig. 1c. The best soils are loams capable of intensive cultivation and of being worked at all seasons. They occur in the Lothians, the Berwickshire Merse and the lowlands from Perth to Stonehaven, where there are major concentrations of arable land (Bibby 1982). These are also the areas having land of the highest capability for farming purposes (Fig. 1d) (Coppock 1980). Patches of the highest-quality land also occur in Ayrshire and along the Clyde estuary. Good, second-quality land is dominant in Buchan, the Central Lowlands and the southwest. This land can support both grass and arable, depending on the prevailing climate. Some second-quality land also occurs in Caithness and the Orkneys. Over three-fifths of all land in Scotland is classified as being of poor quality for farming. Consequently, such land is used only for rough grazing or for forestry and other non-agricultural activities. Table 2 shows the large extent of rough grazing, substantial parts of which are used for field sports, especially deer-stalking and grouse-shooting, rather than for productive agriculture.

In addition to these physical factors, agricultural land use is also affected by the human-made framework of farming, which includes elements such as farm size, tenure, the labour force, machinery and other purchased inputs. In turn this framework reflects the type of farming performed and therefore the detailed pattern of agricultural land use.

The relationship between the human-made framework and the type of farming is well illustrated by the distribution of the farm labour force, which is heavily concentrated in the areas with the most intensive arable and horticultural production, and also in dairying areas. So, as in the case of so many agricultural distributions in Scotland, the main contrasts are between the lowlands and the uplands and, within the uplands, between the Highlands and the Southern Uplands. Of the labour force, 71.4% is full-time compared with 79.0% in 1972. There has been a 39.8% decrease in the labour force since 1972, reflecting one of the many savings farmers have had to make in the face of the rising costs of their inputs.

Holdings are largest in the poorest farming areas, so farms over 400ha dominate in the Highlands and Southern Uplands; these are the main areas of hill sheep-farming, where rough grazings form a substantial part of farm holdings. Holdings of between 120ha and 400ha are most characteristic of eastern and southeast Scotland, which is the main arable area and the leading

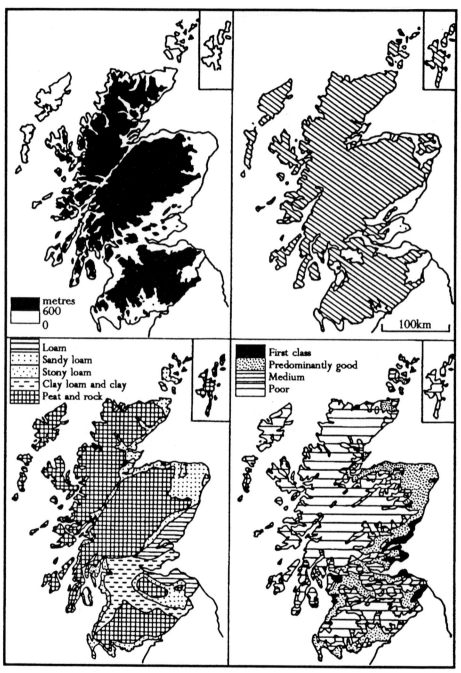

Fig. 1. a. Relief.
       c. Soils.

b. Less Favoured Areas.
d. Land capability.

TABLE 2. Land use in Scotland, 1972 and 1990 (ha) (after DAFS 1973 and SOAFD 1991).

	1972	1990	% Change 1972–90
(1) Total area	7,708,007	7,708,007	
(2) Arable, grassland and rough grazing	6,192,724	5,687,529	–8.16
(2) as a % of (1)	80.34	73.79	–6.55
Rough grazing	4,507,816	3,987,091	–11.55
Rough grazing as a % of (2)	72.79	70.10	–2.69
(3) Crops, fallow and grassland[1]	1,684,908	1,700,438	+0.09
Arable	1,260,295	1,039,158	–17.55
Arable as a % of (3)	74.80	61.11	–13.69
Permanent grassland	424,614	661,280	+55.74
Permanent grassland as a % of (3)	25.20	28.89	+3.69
Labour force (all workers)	46,680	28,108	–39.79
Holdings	36,926	27,166	–26.43
Average holding size[2]	167.7	209.4	+24.87

[1] Throughout this paper, the area of crops, fallow and grassland (excluding rough grazing) is referred to as 'agricultural land'.
[2] For 1972 this consists of: rough grazings 122.1ha, arable 34.1ha, permanent grassland 11.5ha. For 1990 this consists of: rough grazings 146.8ha, arable 38.3ha, permanent grassland 24.3ha.

district for cereal production. An indication of this distribution of small and large holdings is given in Fig. 2, showing the number of holdings per 5km², using data from the annual Agricultural Census. However, this map is slightly misleading as many of the smallest holdings are excluded from the Agricultural Census. This means that many small crofts in the Western Highlands and Islands are not included. In 1990 it was estimated that there were 19,000 such holdings, generally supporting only part-time or spare-time farming activity. These holdings are most numerous in the former crofting counties. As Mather (1983, 208) noted, "in the Highlands, tiny crofts may be juxtaposed with sheep farms extending to thousands of hectares, while in the lowlands, large cropping farms may lie alongside smallholdings created as a result of government policy during the early part of the twentieth century". The elimination of some of these part-time and spare-time holdings from the Agricultural Census in the 1980s accentuates the decline in holdings shown in Tables 2 and 3. It also accentuates the trend towards larger holdings that has occurred during the last two decades and that is illustrated in Table 3. In 1990, 77.8% of holdings were over 20ha, compared with 55.8% in 1972.

Land use

The following description of the distribution of crops and livestock and of changes during the last twenty years is based on data from the annual Agricultural Census. Taken annually since 1866, this collects information from farmers with holdings above a given size and covers crop areas, livestock numbers and the farm labour force. Data from individual holdings are consolidated into

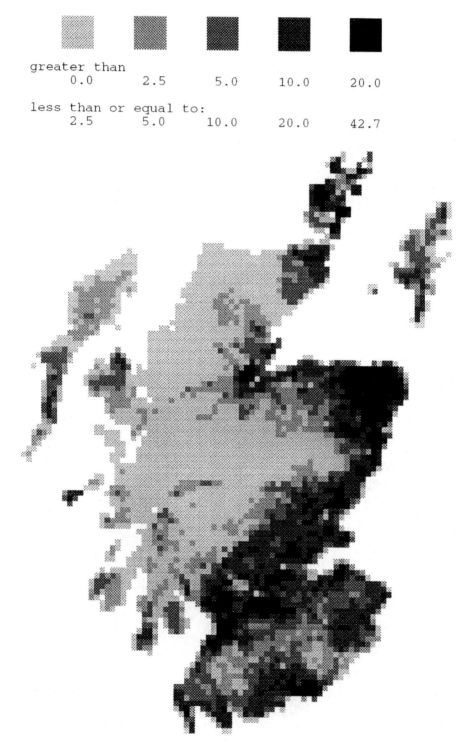

greater than
0.0   2.5   5.0   10.0   20.0

less than or equal to:
2.5   5.0   10.0   20.0   42.7

Fig. 2. Total number of holdings, 1990.

TABLE 3. Farm size structure 1972 and 1990 (after DAFS 1973 and SOAFD 1991).

Size of holdings (ha*)	1972		Size of holdings (ha)	1990	
	Number	%		Number	%
< 2	3,546	9.60	0.1–1.9	792	2.96
2–6	6,073	16.45	2–4.9	1,395	5.21
6–20	6,621	17.93	5–19.9	3,746	14.00
20–40	6,470	17.52	20–49.9	5,600	20.93
40–60.5	4,725	12.80	50–99.9	6,060	22.65
60.5–121	6,220	16.84	100–199.9	4,858	18.16
121–202	2,326	6.30	200–499.9	2,712	10.14
> 202	945	2.56	> 500	1,594	5.95
Total	36,926	100.00	Total	26,757	100.00

*Approximate size categories as original data are given in acreages.

parish summaries and summaries for the regions. Temporal comparisons are restricted by variations in categories and by the lack of application of consistent definitions (Coppock 1978), but data for 1972 and 1990 are compared in Table 4. Despite limitations, the Census does provide the best guide to changes in agricultural land use at a national and regional level. It has also played a major part in many regional studies of agricultural change in Great Britain and its utility in providing data for maps of agricultural distributions is demonstrated here in Figs 2–6.

These maps have been produced using the computer software package GRIDMAP, developed as part of the Economic and Social Research Council's Regional Research Laboratory (Scotland) initiative. This software takes the spatially referenced data of the parish summaries and maps it at a resolution of 1km squares based on the National Grid (Bayley 1988). In examining distributions of different types of land use for the whole of Scotland, the 1km squares have been grouped into 5km blocks. Although there are limitations as regards the degree of accuracy of the maps produced, the salient features of any single type of agricultural land use can be ascertained quickly and easily using this computerised mapping system (Hotson 1988), which is the successor to that employed previously by Coppock (1976).

Grassland

Grass is the most important crop on Scottish farms in terms of area — 62.7% (if rough grazings are excluded). Of this grassland 31.2% is mown. Although permanent grassland occupies a smaller area than arable land, it still accounts for 38.9% of the agricultural land (as defined in footnote 1 of Table 2). In addition, temporary grassland (grassland down for less than five years) occupies 23.8%.

The greatest concentrations of grassland are in the southwest, in conjunction with dairy farming. There it occupies over two-thirds of the agricultural land, though the highest proportions of mown grass occur in the western Highlands

TABLE 4. Agricultural change, 1972–90 (after DAFS 1973 and SOAFD 1991).

| Crop | % of crops, fallow & grass | | % change in hectarage |
	1972	1990	1972–90
Wheat	1.96	6.52	+236.54
Barley	19.72	19.87	+1.67
Oats	5.60	1.74	−68.69
Potatoes	2.17	1.60	−25.62
Turnips/swedes	3.28	1.58	−51.24
Oilseed rape		2.66	
Peas for combining		0.26	
Others for stockfeeding	1.17	0.66	+262.91
Fruit	0.27	0.20	−27.69
Vegetables	0.35	0.65	+89.12
Bulbs, nursery	0.05	0.01	−3.36
All others	0.03	0.12	+390.44
Bare fallow	0.22	1.41	+549.71
Tillage	34.81	37.34	+8.26
Temporary grass	39.99	23.77	−40.01
Permanent grass	25.20	38.89	+55.74
Grassland	65.19	62.66	−3.00
Grass for mowing	19.39	19.55	+1.79

| | Density per 100 ha crops, fallow & grass | | % change in numbers |
	1972	1990	1972–90
Cattle	141.71	122.68	−12.63
Sheep	448.20	552.85	+24.49
Pigs	39.28	26.40	−32.16
Poultry	837.33	858.25	+3.44
Labour	2.77	1.65	−39.79

and Islands. These are the areas with the least tillage and where cut grass forms a significant component of winter feed. Permanent grass is dominant in areas least suitable for cropping and on the margins of cultivation, and hence its concentrations in the western Highlands and Islands and inland parts of the southwest. Temporary grassland has a more even distribution than permanent grassland and is an important feature of farming in the lowlands, especially in the northeast where it accounts for 90% of the grassland. Here it has formed part of a traditional crop rotation also involving oats and turnips and swedes.

Tillage

Tillage, i.e. land under fallow or crops other than grass, represents just over one-third of Scotland's agricultural land, and the sale of crops accounts for one-quarter of the country's agricultural output; in addition, large quantities of crops are consumed by livestock. Tillage reaches its greatest proportion of the agricultural area in eastern Scotland, from the Berwickshire Merse to the Moray Firth, in areas where soils and climate provide the best conditions for crop production. Over half the agricultural land is in tillage in the Merse, the Lothians, eastern Fife and the coastal lowlands of Tayside. Most of the area used for tillage is under cereals.

Cereals

"Scottish-bred cereals are very tolerant of cool summer temperatures. They mature late and hence can take advantage of the whole growing season" (Tivy 1983, 69–70). Cereals accounted for two-thirds of the sales of crops in 1991, the greatest revenue coming from barley, which dominated the cereal acreage. The barley-growing areas are in some of the best-quality land in the country in eastern Scotland (Fig. 3a). The crop is grown for seed, for malting and for livestock feed, one-third being retained on the farms on which it was grown for this latter purpose. Its prominence in the tilled area is greatest in the Merse, Buchan and around the Cromarty Firth. The area under the crop has expanded considerably in the post-war period. This occurred initially through its use for cattle feed in so-called 'barley beef' production. In this system cattle are fed on a concentrated cereal diet with little roughage, gaining weight quickly to be ready for slaughter within twelve months of birth and producing lean meat favoured by the market. More recently, favourable guaranteed prices within the CAP have encouraged both wheat and barley production (Dawson 1980).

Wheat is the second most widely grown cereal, having displaced the traditional Scottish cereal, oats, from this position since 1972. The area under wheat has more than trebled in response to the favourable subsidies offered by the CAP. Compared with barley it has a more confined distribution, with a concentration in Lothian and Fife, reflecting the more favourable soils and climate there (principally the higher temperatures and drier conditions) (Fig. 3b). As well as the growth in the areas under wheat and barley, the production of both crops has also increased through higher yields. Wheat yields, for example, have doubled since the 1950s and illustrate the tremendous technical developments that have led to substantial growth of output from the same unit input of land.

Oats now occupy less than 2% of agricultural land, though the crop is grown in small quantities throughout the country. The area under oats has declined since 1942, reflecting lower yields than from other cereals, reduced demand because of the replacement of working horses on farms, and the increased substitution of barley for oats as a livestock feed. Moreover, oats are not very suitable for harvesting by combine harvesters as the grain can leave the seedhead too easily if it is over-ripe. Oats are still an important part of crop rotations in Buchan, and are grown on many farms in small hectarages, even in the north and west in areas unsuitable for other cereals.

Root crops

Three of the traditional staples of Scottish agriculture, oats, potatoes and turnips/swedes, have declined significantly since 1972 as crop rotations involving oats and root crops have ceased to be of great importance. Turnips and swedes have long been important fodder crops, largely used for feeding to sheep on the

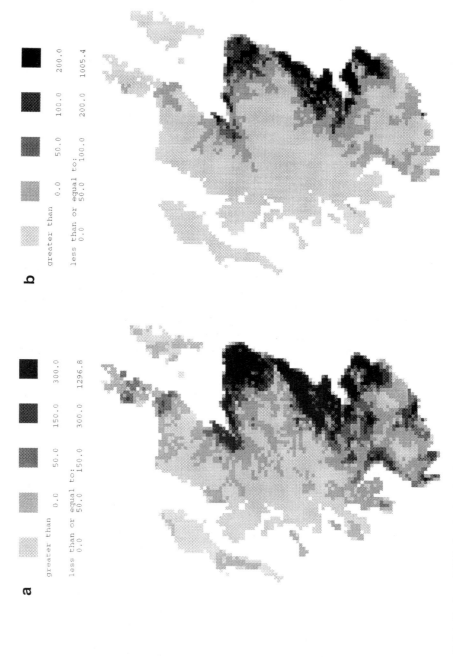

Fig. 3 (a) Barley, 1990 (includes both winter and spring barley) and (b) Wheat, 1990.

farms on which they are grown. However, since 1945 the decline of the arable sheep flock has accelerated the decline of fodder roots, though turnips are still used for fattening Cheviot lambs (Robinson 1988, 232).

Potatoes were the leading cash crop in Scotland in the 1960s, but have now been displaced in importance by barley. The area under potatoes has fallen by one-quarter since 1972, though production has not declined. Yields have increased whilst the area under the crop has been regulated by the Potato Marketing Board, with which growers have to register. The main concentrations of potatoes are in the eastern half of the Central Lowlands, especially Tayside. They are also the main root crop in the western Highlands and Islands, where there is still some subsistence production in the crofting areas. Early potatoes are most important in the southwest. The cool and windy summer weather limits insect pests and pathogenic organisms so that Scotland is also a major producer of disease-free seed potatoes (Tivy 1983, 71).

Oilseed rape

In just ten years oilseed rape has become a more widely grown crop than many of the traditional mainstays of Scottish farming (Fig. 4). Given the deficiency in vegetable oils and protein meal in western Europe, the CAP established attractive prices to encourage farmers to produce oilseed rape, as the rapeseed, when processed, gives an edible oil suitable for salad and cooking oils and margarine. A by-product, rapemeal, is a useful protein concentrate in animal feedstuffs. New varieties have helped to produce less acidic animal feed and to accelerate the crop's spread in the UK (Wrathall 1978). First adopted on any scale in the south of England just prior to entry to the EC, the sight of the yellow fields of ripe rapeseed has become common in spring in many parts of the country. This spread has reflected the establishment of new crushing plants in Glasgow and northern England and the lower incidence of pests and diseases affecting rape in cooler conditions (Bunting 1984).

By 1983 the area under oilseed rape in the UK exceeded that of sugar-beet and potatoes, so that it had become the largest non-cereal arable crop in the country, with sales exceeding £200 million (Wrathall and Moore 1986, 352). By this time oilseed processors as well as growers were receiving subsidies, and for growers the gross margins per unit area were 20% above those for grains of feed quality.

The crop was first grown in Scotland in Berwickshire in 1979–80, with farmers sowing rape in winter on low-lying land in the east of the county. By 1985–6 the number of growers in the county had risen to 71 (Gillie 1987), and there were small areas of oilseed rape being grown as far north as Caithness and in the Western Isles (Robinson 1983, 193–7). In 1990 Scotland had over 45,000ha under the crop, the main centres of production being the Merse, Lothian, Fife, Tayside, Buchan and around the Cromarty Firth. Small amounts were also grown for livestock feed on well-drained land in western Scotland.

Horticulture

Table 5 shows clearly the two contrasting experiences of horticultural production during the 1970s and 1980s: a reduction of the area under fruit crops but an almost doubling of that under vegetables. The latter reflects a reduction in the reliance of Scottish consumers upon vegetables grown in England. In addition, there are now more field crops of vegetables being grown in Scotland

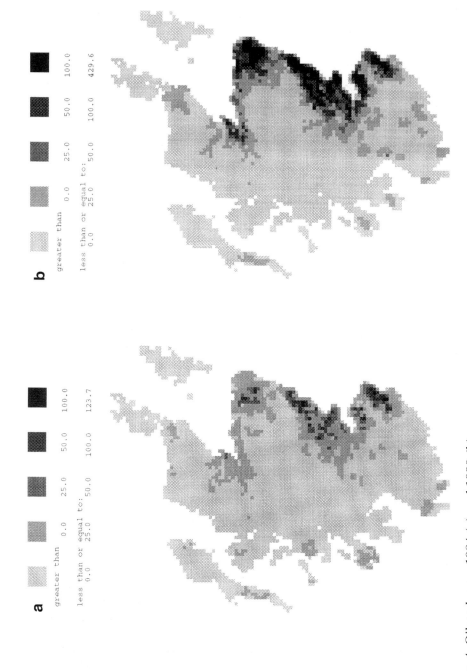

Fig. 4. Oilseed rape, 1984 (a) and 1990 (b).

TABLE 5. Changes in the horticultural area, 1972–90 (after DAFS 1973 and SOAFD 1991).

Crop	Hectares 1972	1990	% Changes 1972–90
(a) *Vegetables for human consumption*			
Brussels sprouts	536	232	−56.7
Cabbages/calabrese	784	1,459	+86.1
Cauliflowers	361	712	+97.2
Carrots	788	993	+26.0
Peas	1,812	4,414	+143.6
Turnips and swedes	421	1,881	+346.8
Leeks	155	177	+14.2
Lettuces (not under glass)	195	265	+35.9
Rhubarb	203	87	−57.1
Beans	na	587	
Beetroot	155	na	
Other vegetables	342	279	−18.4
Glasshouses	110	na	
All vegetables	5,862	11,086	+89.1
(b) *Nursery crops*			
Hardy nursery stock	424	385	−9.2
Cut flowers	82	17	−79.3
Bulbs	297	333	+12.1
Other nursery crops	0	42	
All nursery crops	803	777	−3.2
(c) *Small fruit*			
Strawberries	976	649	−33.5
Blackcurrants	44	256	+481.8
Raspberries	3,368	2,316	−31.2
Others/mixed fruit	209	60	−71.3
All small fruit	4,597	3,281	−28.6
(d) *Orchards*			
All orchards	143	43	−69.9
Total horticultural area	11,405	15,187	+33.2

on contract to local processors, wholesalers and retailers. The increased area of vegetables grown for human consumption has largely comprised peas (in Tayside) (see Leveson 1981), cauliflowers (East Lothian), turnips and swedes (Lothian), and cabbages (East Lothian). Midlothian and East Lothian have remained the principal area for vegetable-growing, with a secondary concentration in the lowlands between Perth and Stonehaven. The latter area is also the main one for production of soft fruit, especially the main soft fruit crop, raspberries. The cool summers, with a critical dry period at harvest time, favour higher yields and quality of this crop in Scotland than in England. However, competition from producers in the EC and that from a wider range of international fruit readily available in Scottish supermarkets has cut the area under fruit in Scotland by over one-quarter since 1972. Raspberry-growers are suffering from cheap imports from eastern Europe and a shortage of suitable processing outlets (Daw and Wright 1991). Commercial top-fruit orchards have become more concentrated in the lower Clyde Valley in Lanarkshire.

The other sector of the horticultural industry is nursery crops, which account for just 5.1% of the area under horticulture. Approximately half of this area of nursery crops is under bulbs, with the remainder under hardy nursery stock. The 1986 Survey of the Hardy Nursery Stock Industry recorded nearly 300ha of commercial operations, with an average of 2.74ha per operation. The total value of sales was just over £7 million gross and £4.9 million net.

Woodland

The area under farm woodlands has increased in recent years, partly through the operation of initiatives to promote farm forestry. These have included the Farm Woodland Scheme and the Woodland Grant Scheme, both aimed at establishing woodland as a 'normal' part of commercial agriculture through relatively small-scale grant encouragement. Uptake has been greatest in Grampian Region and generally on eastern arable farms (Robinson 1991a, 100). Farmers taking part in the schemes have had multiple objectives in mind: creating new woodlands for environmental and amenity reasons, for purposes of conservation, as shelter-belts, and for game management (Scambler 1989; Appleton 1990). Prior to the start of these schemes, the Farm Woodlands Survey of 1985–6 estimated the area of farm woodland at 56,000ha, spread over 48,000 holdings, or 27% of holdings in Scotland. In addition there were 22,000ha of tree-dominated scrub on 3500 holdings. On holdings with woodland the average wooded area was 6.7ha. Southeast and southwest Scotland accounted for nearly three-quarters of this farm woodland. Its main uses were as shelter for livestock, general amenity or landscape value, and the conservation of woodland (Dellaquaglia 1987).

Farm livestock

Cattle

The numbers of cattle in Scotland have fallen by over 12% since the UK joined the EC, though the sale of fat cattle and calves remains the chief single item of revenue to Scottish farming and the sale of milk and milk products is second. Three-fifths of all Scottish farms have some cattle, with concentrations occurring in the northeast and south-west. The highest densities of beef cattle are reached in Buchan and Orkney but, as a percentage of all cattle, beef are dominant nearly everywhere except in the Central Lowlands and the main dairying region of the southwest (Fig. 5a). Beef cattle are most important in the economy of the northeast and the Orkneys. The former represents the main area in Scotland for the fattening of cattle, which are transferred there from upland rearing-grounds in the west and from the Orkneys. The northeast is the home of the Aberdeen Angus breed, specially developed to provide substantial amounts of lean beef. Cross-breeding of the Aberdeen Angus with Beef Shorthorns, Galloways and Herefords is also common.

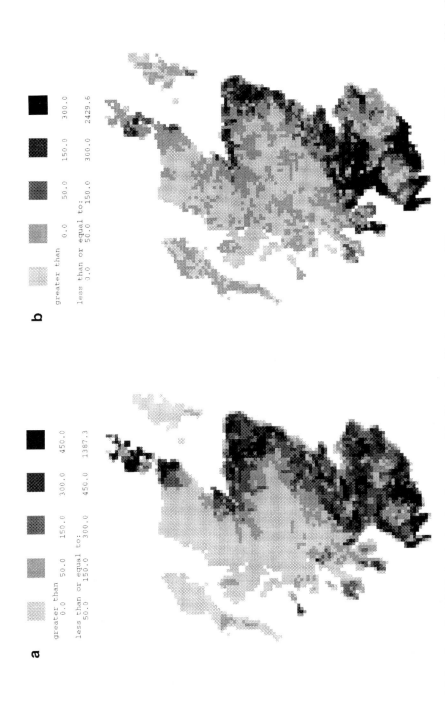

Fig. 5. (a) Beef cattle, 1990 (includes beef cows and heifers in milk, beef cows in calf but not in milk, beef heifers in calf for the first time, and beef females for breeding); (b) Dairy cattle, 1990 (includes dairy cows and heifers in milk, dairy cows in calf but not in milk, dairy heifers in calf for the first time, and dairy females for breeding).

The other well-known Scottish cattle breed is the Ayrshire, the dominant dairy breed in Scotland, which has its highest concentrations in the main dairying areas of the Ayrshire lowlands, the Clyde Valley and the coastal lowlands along the Solway Firth. Dairy cattle are also found in the eastern half of the Central Lowlands and around Aberdeen (Fig. 5b), but much of Scotland does not possess conditions conducive to dairying. Dairying is primarily a lowland activity, strongly associated with the main grassland areas, though many farms have a few 'house cows' producing milk for the farm.

The numbers of dairy cattle have fallen by 9% since the EC introduced a quota system for milk production in 1984. This was the first of a series of measures introduced to curb over-production of agricultural produce in the EC, but the further use of quotas was eschewed because of operational and administrative difficulties (Groves 1989). Quotas have led to reduced profits on dairy farms and have contributed directly both to a reduction in the number of milk producers and an increase in average herd size (Cook and Smith 1987; Naylor 1990).

Sheep

In contrast to cattle, the numbers of sheep have risen by nearly one-quarter since entry to the EC, reflecting the continued support given to hill farming via the Compensatory Allowances of the Less Favoured Areas scheme (see below). This has tended to favour increased stocking densities of both sheep and beef cattle, especially in the Highlands and the Southern Uplands. Sheep have continued to be most numerous in the Southern Uplands and Tweed Basin, though their distribution is a wide one and sheep are kept on nearly half of all Scottish farms (Fig. 6a). They are most numerous on hill farms, utilising extensive tracts of rough grazing land which cannot support other farming enterprises. However, there has been a contraction of the area under rough grazings, reflecting the spread of afforestation.

Different breeds dominate different parts of the country and there is a complex pattern of movements for breeding, fattening, marketing and slaughter, as shown in Fig. 6b. Pure-bred Blackface and Cheviots are maintained in harsh, 'high' uplands. These flocks supply female lambs for cross-bred flocks in less harsh 'low' uplands. Here the Blackface and Cheviot ewes are crossed with Border Leicester rams to produce Greyface and Half-Bred lambs respectively. In turn, these flocks provide female breeding replacements to lowland cross-bred flocks where Half-Bred and Greyface ewes are crossed with Down rams to produce Down-cross lambs. This pattern of movement necessitates replacement of one-quarter to one-third of the total breeding stock annually. Markets for sheep tend to transfer sheep from the north and west to the east and south (Carlyle 1972a; 1972b; 1975; 1978).

Pigs and poultry

Pigs and poultry accounted for 9% of agricultural output in 1990 and were concentrated on much fewer holdings than sheep and cattle. Both pigs and poultry have patchy distributions: pigs are concentrated around the major cities, especially Edinburgh, and poultry in the central and eastern lowlands. Competition from other EC producers, notably Denmark, has helped to reduce pig numbers by one-third since 1972. This reduction has also concentrated production more into areas of intensive arable production. Hence pig-keeping

(a)

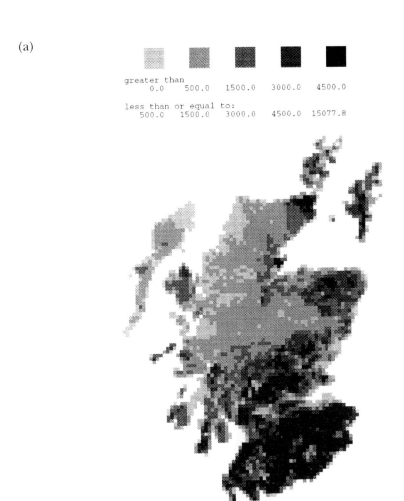

(b)

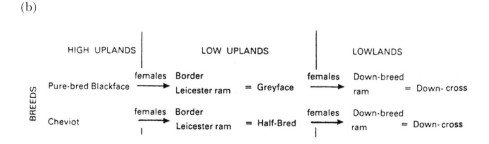

Fig. 6 (a) Sheep, 1990.
   (b) The distribution and movements of sheep breeds in Scotland (after Carlyle 1972a).

has been a way of maximising profits from barley-growing by using barley in pig-feed (Symes and Marsden 1985).

Poultry numbers have changed little since the 1960s, partly because increased egg production per bird has enabled market demand to be satisfied. The reduction in egg-laying birds has been balanced by greater production of table fowls, including broilers. Much of this rising output of poultry meat has been produced on contract for large retailers or processors. This has fostered concentrations of production near processing factories close to the main centres of population. In addition to these high concentrations, focusing upon industrial-style production of eggs and poultry meat for market, many farms in Scotland maintain small flocks for domestic purposes.

Agriculture under the Common Agricultural Policy (CAP)

The preceding descriptions of the pattern of agricultural land use in Scotland have also highlighted some of the changes occurring since the UK's entry into the EC in 1973. There have been some strong responses to the stimuli of the CAP's price support system, as well as continuing reactions to consumer demand. The increased importance of a few large retailers and wholesalers has effected substantial changes in the marketing of farm produce, but the major changes in land use have reflected the operation of the CAP. Its price guarantees and its guidance schemes have created a new climate for farming, replacing the previous mechanisms of support established in the 1947 Agriculture Act. However, from the early 1980s growing recognition of weaknesses in the CAP and of undesirable consequences arising from its implementation have produced a number of reforms. These reforms and those promised during the 1990s will further continue changes in agricultural land use before the end of the century.

In the mid-1970s encouragement was given to farmers in upland areas by way of the Less Favoured Areas (LFAs) scheme. This perpetuated one of the basic aims of the CAP, namely the maintenance of farming communities throughout the EC. Subsequently, to this type of support for farmers facing difficult physical conditions were added measures to improve the structure and general condition of farming. In Scotland the two best examples of this are the Integrated Development Programme for the Western Isles and the Agricultural Development Programme for Scottish Islands (Revell 1990). From the mid-1980s, though, the general tenor of CAP policy initiatives has changed in favour of reforms to curtail production and to encourage less intensive farming practices. One of these reform measures, creating Environmentally Sensitive Areas (ESAs), has been taken up enthusiastically in Scotland. Others, such as taking arable land out of production (set-aside) and promoting the establishment of new on-farm enterprises, have been more widely adopted in southern England (Ilbery 1990; Ilbery and Stiell 1991; Robinson 1991b).

Less Favoured Areas (LFAs)

LFAs were established by the EC in 1975 as part of the declared CAP guidance programme to improve the life of farmers in areas of environmental difficulty (Arkleton Trust 1982). LFAs were defined as areas in danger of depopulation and where maintenance of viable farming communities was seen as essential to the continuing development of the rural economy. In particular, three characteristics were highlighted:

(a) the presence of infertile land, unsuitable for cultivation or intensification, with a limited potential which cannot be increased except at excessive cost, and mainly suitable for extensive livestock farming;

(b) because of this low productivity of the environment, results which are appreciably lower than the mean with regard to the main indices characterising the economic situation in agriculture;

(c) either a low or a dwindling population, predominantly dependent on agricultural activity, and the accelerated decline of which would jeopardise the viability of the area concerned and its continued habitation.

By applying specific criteria (see Robinson 1988, 219), this initially led to the establishment of LFAs covering 41% of the surface area of the UK. Further additions to this area in 1982 and 1985 extended the coverage to 53% of the UK, 65% of which is in Scotland. In fact, LFA status has been conferred on over three-quarters of Scotland, entitling farmers to compensation for the permanent natural handicaps with which they have to contend. Special payments, geared mainly to maintaining beef cattle herds and sheep flocks, are intended to ensure the continuation of farming whilst encouraging farm modernisation, maintenance of a minimum population level and conservation of the countryside (Weinschenk and Kemper 1981).

The compensation measures have tended to benefit larger farms disproportionately and so have not realised the social objectives of the policy. Furthermore, the assumption that maintenance of stock-rearing in the uplands is the best way of preserving and enhancing the environment has been questioned (Bowler 1985, 210). Recognition of these and other defects in the LFAs scheme is now producing policy revisions placing less emphasis on retaining existing stocking rates and more on retaining the appearance of existing landscapes.

Development programmes for disadvantaged areas
Funded by both the EC and the UK government, the Integrated Development Programme (IDP) for the Western Isles was intended to promote socio-economic development in a region where agriculture alone has been unable to sustain economic growth (Nurminen and Robinson 1985; Robinson 1988, 222–7). Of the farm holdings in the Isles, 89% are under 20ha, emphasising the significance of part-time smallholdings or crofts. Not surprisingly, the IDP aimed to provide additional employment opportunities for part-time farmers whilst also improving the structure of farming itself. Particular targets were the need to improve the quality of pasture, to improve the quality of livestock, to promote desirable farming practices through grants for drainage, fencing and fertiliser application, and to promote improved livestock marketing through cooperative ventures. Between 1982 and 1987, over £28 million was spent on agriculture and fish farming in the IDP, with nearly 90% of crofters participating in at least one of the IDP schemes. Approximately one-quarter of this investment remained in the local economy to boost incomes. It also had the effect of halting the decline in cattle numbers that had occurred in the 1970s. It produced a 10% rise in the size of the Isles' sheep flock, and a sevenfold increase in sheep 'exported' from the Isles. The latter has been greatly assisted by investment in improved

communications, including both up-grading of roads and improvement of ferry services (Houston 1987).

The Agricultural Development Progamme (ADP) for Scottish Islands was launched in 1988 with five years of funding from the Highlands and Islands Development Board (HIDB). Founded in 1965, the Board's remit was to improve economic and social conditions in Scotland's northernmost region. To this end, until its recent replacement by Highlands and Islands Enterprise, it operated as a stimulator of new private enterprises especially in fishing, craft industry and tourism. In the ADP, though, the focus was upon a number of objectives similar to those in the IDP. These included improving farm operations and efficiency, encouraging greater production of winter feed, increasing the amount and quality of farm output, promoting diversification into non-farming activities, and promoting environmental conservation. Grant incentives have been provided to stimulate a range of improvements, with the highest premiums attached to land improvement and increasing the quality of farm livestock. A total of £38 million has been committed to the ADP.

Environmentally Sensitive Areas (ESAs)

ESAs were introduced in the UK under the 1986 Agriculture Act, which followed an EC Regulation of 1985 permitting member states to introduce an aid scheme "to contribute to the introduction or maintenance of farming practices compatible with the requirements of the protection of the environment", and particularly in areas very sensitive from this point of view.

The UK legislation enables designation of an area as an ESA if it is particularly desirable to protect its natural beauty, flora or fauna, or historic or archaeological features, and if this objective is likely to be assisted by the maintenance or adoption of particular agricultural methods. This last objective accounts for the strong relationship between farming and environment in ESAs. In an ESA, farmers can enter into agreement with the secretary of state to receive payments in return for undertaking farming practices beneficial to the environment.

In Scotland the Breadalbane and Loch Lomond ESAs were introduced in 1987, with a two-tier system of payment to farmers: flat-rate payments per ha for agreeing to follow specified prescriptions; and additional specific payments related to particular items of conservation work, e.g. fencing, dyking or hedging. The details of the agreed scheme are set out in a conservation plan covering the whole of a farm. Thus ESAs place emphasis upon positive conservation measures as well as basic protection from agricultural intensification (DAFS 1989).

Three more ESAs were established in Scotland in 1988 (Fig. 7) — in the Stewartry, Whitlaw and Eildon, and the Machair of the Western Isles. By early 1992, 811 farmers and crofters had joined the scheme in the five areas, representing 65% of those farmers eligible to join and 121,000ha or 70% of the eligible land. Nearly £3 million had been paid out under the scheme by this time.

Monitoring of the first two ESAs was carried out in 1991, resulting in revised financial incentives and changes to boundaries and administration, including a wider range of measures for the enhancement of woodlands, wetlands, herb-rich pastures, heather regeneration and archaeological sites. Subsequently, more ESAs have been designated in which a prime objective will be regeneration of damaged heather and other forms of upland vegetation. The new ESAs will be in the Argyll Islands, Shetland, the Cairngorm Straths, the Western Southern

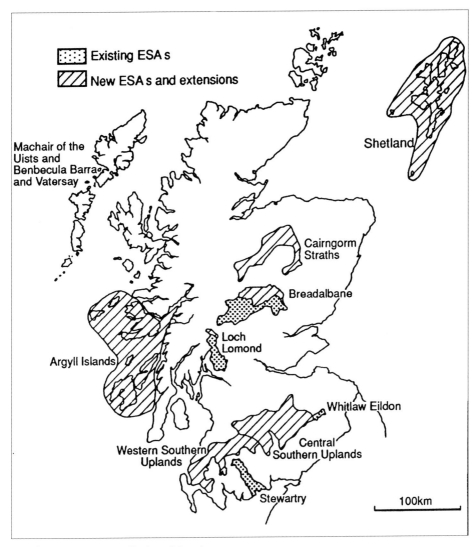

Fig. 7. Environmentally Sensitive Areas.

Uplands and the Central Southern Uplands. The new ESAs and an expansion programme for Breadalbane and Loch Lomond will account for £12 million of new expenditure, giving over 3000 more farmers the opportunity of joining the scheme and extending ESA coverage to around 15% of the total land area of Scotland.

Conclusion

The measures referred to above suggest that farming in Scotland is entering a new phase, as European agriculture in general moves away from a 'productionist era', maintained by substantial price guarantees, to a 'post-productionist era' in which farmers are paid 'to produce countryside'. Recent proposals to modify the CAP suggest cuts in cereal prices to move into line with world prices, with both reductions in the cereal acreage and promotion of more extensive farming practices. This could have a significant impact on lowland eastern Scotland, with more farmers forced to turn to new sources of income, and perhaps increasingly from off-farm activities.

The uncertainty being caused by changes in agricultural policy comes at a time when many farm incomes are greatly reduced from their level of ten years ago and when continued loss of full-time farm labour is occurring (Britton 1990; Hill 1990). So the turmoil of unfolding events offers the prospect of sustained opposition from many farmers and further political bargaining as EC member countries seek to maximise their own interests. For Scotland an extra dimension to the proposed reforms is provided by the large extent of the country that is in receipt of grants because of the unfavourable physical conditions for agriculture. Will farmers in such areas continue to receive compensatory payments to help them produce livestock, or will they be paid to cut back on production whilst maintaining attractive farmed landscapes?

There is a strong possibility that clearer regional divisions will develop under the CAP reforms. Large, efficient arable farms in the eastern lowlands seem best able to gain advantage from the reforms, whilst small, marginal livestock producers in the uplands and small dairy units will continue to struggle. It is likely that the flight from the land by both farmers and labourers will continue and that concentration of farm ownership will accompany the continued growth in holding size. Farming in AD 2000 will therefore be leaner, but possibly facing continued confusion as to what goals are to be attained in an EC in which over- rather than under-production is dominating thinking on agricultural policy.

Acknowledgements

I am grateful to Jack Hotson and Alison Bayley of the Data Library, University of Edinburgh, for helping to produce Figs 2–5 and 6a, and to Anona Lyons for drawing Figs 1, 6b and 7.

References

Appleton, Z.E.N. 1990 The impact of the Farm Woodland Scheme in Scotland. *Scottish Agricultural Economics Review* **5**, 145–57.

Arkleton Trust 1982 *Schemes of assistance to farmers in Less Favoured Areas of the EEC.* Langholm. Arkleton Trust.

Bayley, A. 1988 A GRIDMAP user guide. Working Papers, Regional Research Laboratory for Scotland. No. 4.

Bibby, J.S. (ed.) 1982 *Land capability classification for agriculture.* Aberdeen. Macaulay Institute for Soil Research.

Birse, E.L. and Dry, F.T. 1969 *Assessment of climatic conditions in Scotland. Volume 1.* Aberdeen. Macaulay Institute for Soil Research.

Bowler, I.R. 1985 *Agriculture under the Common Agricultural Policy: a geography.* Manchester University Press.

Britton, D. 1990 Recent changes and current trends. In D. Britton (ed.), *Agriculture in Britain: changing pressures and policies,* 1–33. Wallingford. CAB International.

Bunting, G.S. 1984 Oilseed rape in perspective: with particular reference to crop expansion and distribution in England 1973–1983. *Aspects of Applied Biology* **6**, 11–21.

Carlyle, W.J. 1972a The marketing and movement of Scottish hill lambs. *Geography* **57**, 10–17.

Carlyle, W.J. 1972b The away wintering of ewe hoggs from Scottish hill farms. *Scottish Geographical Magazine* **88**, 100–14.

Carlyle, W.J. 1975 Livestock markets in Scotland. *Annals of the Association of American Geographers* **65**, 449–60.

Carlyle, W.J. 1978 The distribution of store sheep from markets in Scotland. *Transactions of the Institute of British Geographers,* new series, **31,** 226–45.

Cook, P. and Smith, J. 1987 The effects of milk quota cuts on dairy farming in the Scottish Highlands and Islands. *Scottish Agricultural Economics Review* **2**, 57–68.

Coppock, J.T. 1976 *An agricultural atlas of Scotland.* Edinburgh. John Donald.

Coppock, J.T. 1978 Land use. *Reviews of UK Statistical Sources* **14** (8), 1–101.

Coppock, J.T. 1980 The concept of land quality: an overview. In M.F. Thomas and J.T. Coppock (eds), *Land assessment in Scotland,* 1–8. Aberdeen University Press.

DAFS 1973 *Agricultural Statistics 1972 Scotland.* Edinburgh. HMSO.

DAFS 1989 *Environmentally Sensitive Areas in Scotland: a first report.* Edinburgh. Department of Agriculture and Fisheries for Scotland.

Daw, M.E. and Wright, E. 1991 Background to the crisis in Scottish raspberry production. *Scottish Agricultural Economics Review* **6**, 149–64.

Dawson, A.H. 1980 The great increase in barley growing in Scotland. *Geography* **65**, 213–17.

Dellaquaglia, A. 1987 Farm woodlands. In DAFS, *Economic Report on Scottish Agriculture 1986,* 17–19. Edinburgh. Department of Agriculture and Fisheries for Scotland.

Gillie, G.C. 1987 Oilseed rape: a geographical perspective. Unpublished B.Sc. dissertation, Department of Geography, University of Edinburgh.

Groves, C.R. 1989 The impact of quotas on Scottish milk production and processing industries. In A. Burrell (ed.), *Milk quotas in the EC,* 89–99. Wallingford. CAB International.

Hill, B. 1990 Incomes and wealth. In D. Britton (ed.), *Agriculture in Britain: changing pressures and policies,* 135–60. Wallingford. CAB International.

Hotson, J. McG. 1988 Land use and agricultural activity: an areal approach for harnessing the Agricultural Census of Scotland. Working Papers, Regional Research Laboratory for Scotland, No. 11.

Houston, G. 1987 Assessing the IDP for the Western Isles. *Scottish Geographical Magazine* **103**, 163–5.

Ilbery, B.W. 1990 Adoption of the arable set-aside scheme in England. *Geography* **76**, 69–73.

Ilbery, B.W. and Stiell, B. 1991 Uptake of the Farm Diversification Grant Scheme in England. *Geography* **76**, 259–63.

Leveson, I.D.H. 1981 An account of the spread of pea production and pea processing in eastern Scotland, 1969–79. Unpublished M.A. thesis, Department of Geography, University of Edinburgh.

Mather, A.S. 1983 Rural land use. In C.M. Clapperton (ed.), *Scotland: a new study,* 196–223. Newton Abbot. David and Charles.

Miller, R. 1973 Bioclimatic characteristics. In J. Tivy (ed.), *The organic resources of Scotland: their nature and evaluation,* 12–23. Edinburgh. Oliver and Boyd.

Naylor, E.L. 1990 Quota mobility and the changing structure of milk production in north-east Scotland. *Scottish Agricultural Economics Review* **5**, 77–90.

Nurminen, E. and Robinson, G.M. 1985 Demographic changes and planning initiatives in Scotland's Northern and Western Isles. Research Discussion Papers, Department of Geography, University of Edinburgh, No. 20.

Revell, B.J. 1990 Review of support policies for fragile rural areas. *Scottish Agricultural Economics Review* **5**, 109–34.

Robinson, G.M. 1988 *Agricultural change: geographical studies of British agriculture.* Edinburgh. North British Publishing.

Robinson, G.M. 1990 *Conflict and change in the countryside: rural society, economy and planning in the Developed World.* London and New York. Belhaven Press.

Robinson, G.M. 1991a The environment and agricultural policy in the European Community: land use implications in the United Kingdom. *Land Use Policy* **8**, 95–107.

Robinson, G.M. 1991b Changing policies for the management of rural resources in the United Kingdom. *Scottish Association of Geography Teachers Journal* **20**, 29–39.

Robinson, G.M. 1991c The environmental dimensions of recent agricultural policy in the United Kingdom. In G.E. Jones and G.M. Robinson (eds), *Land use change and the environment in the European Community*, 63–76. Biogeography Monographs, No. 4. London. Institute of British Geographers.

Scambler, A. 1989 Farmers' attitudes towards forestry. *Scottish Geographical Magazine* **105**, 47–9.

Scottish Office Department of Agriculture and Fisheries (SOAFD) 1991 *Economic Report on Scottish Agriculture, 1990.* Edinburgh. HMSO.

Symes, D.G. and Marsden, T.K. 1985 Industrialization of agriculture: intensive livestock farming in Humberside. In M.J. Healey and B.W. Ilbery (eds), *The industrialization of the countryside*, 99–120, Norwich. Geo Books.

Tivy, J. 1983 The bio-climate. In C.M. Clapperton (ed.), *Scotland: a new study*, 64–93. Newton Abbot. David and Charles.

Weinschenk, G. and Kemper, J. 1981 Agricultural policies and their regional impact in Western Europe. *European Review of Agricultural Economics* **8**, 251–81.

Wrathall, J.E. 1978 The oilseed rape revolution in England and Wales. *Geography* **63**, 42–5.

Wrathall, J.E. and Moore, R. 1986 Oilseed rape in Great Britain — the end of a 'revolution'. *Geography* **71**, 351–6.

In: A. Fenton and D.A. Gillmor (eds) 1994 *Rural land use on the Atlantic periphery of Europe: Scotland and Ireland*, 99–116. Dublin. Royal Irish Academy.

AGRICULTURE AS A LAND USE IN IRELAND

James A. Walsh

Abstract: Agriculture is by far the most important land use in Ireland. While the fundamental division between field crops and grassland-based enterprises has altered very little over the past two decades, there have been significant shifts in emphasis in relation to the use of grassland, mainly in response to changes in the operation of the Common Agricultural Policy (CAP). There has been a reduction in the size of the dairying herd, which has been largely compensated for by increases in the numbers of suckler cows. By far the most important change has been the expansion in sheep-rearing in response to the establishment of a common market for sheep meat in 1980. Despite considerable differences between regions in the composition of their livestock sectors, there was relatively little variation in the overall levels of intensification, except in the southwest. The most important land-use implication of the 1992 reforms to the CAP probably relates to the set-aside proposal for land used to grow cereals, since it will apply to some of the best agricultural land.

Introduction

Agriculture is Ireland's largest single industry. It accounts for approximately 10% of Gross Domestic Product (GDP) and 14.5% of employment in the Republic of Ireland, while in Northern Ireland the comparable figures are 4.5% and 7.0%. Land is the principal component of the agricultural production system. The transformation of land resources into land uses is influenced by several factors, related to processes that may be broadly categorised as physical, historical, behavioural, economic and political. The high level of geographical variability in the quality and potential of agricultural land has been noted by Lee (this volume). Social patterns of ownership and control of land have developed over lengthy periods and are today reflected in variable farm sizes, field patterns and levels of fragmentation. For example, in relation to farm size there is a significant contrast between the eastern and southern parts of the country, where farms tend to be relatively large, and some northwestern and western districts where the majority of farms are less than 20ha in size. The difference in farm sizes also coincides in part with the variability in land quality, so that many of the small farms are in areas of low agricultural potential (Gardiner and Radford 1980). Another important historical influence is the accumulated tradition, skills and attitudes concerning particular land uses which have been built up in

99

different regions over many decades. This legacy may be either open and adaptive or closed and averse to risk, which may in turn influence the patterns of adjustment to changing economic circumstances. The goals, values and expectations of farmers are also important. Very little research has been undertaken on the importance of behavioural factors in decisions relating to agricultural land use. Gillmor (1986) has provided some tentative evidence, which confirms findings from studies in other areas, that farmers are motivated by a variety of goals, and that the rank ordering of goals may vary between different categories of farmers in different regions. Individual behavioural factors are also relevant in that they influence decisions on investments that may result in improvements in the quality of some land resources and increase their range of potential uses, e.g. investment in drainage or the use of fertilisers.

The most important influences on recent changes in agricultural production, however, have come from beyond the immediate environment of the farmer and the farm. They are related to changing international market conditions which have also resulted in major changes in European Community (EC) policy towards agriculture. Since the Second World War there has been a strong political commitment to supporting agriculture, especially the grassland-based enterprises in which Ireland has a competitive advantage (Boyle 1992). This is an important factor given the high level of agricultural exports from both parts of Ireland. From the early 1970s the main influence has been the Common Agricultural Policy (CAP) of the EC. Under the CAP substantial price supports have been paid to farmers in order to stimulate increased output. Each of the main sectors in Irish agriculture — dairying beef, cereals and sheep meat — has benefited under this regime.

Throughout the 1970s, substantial increases in agricultural production in Ireland and the EC were associated with greater specialisation in farming systems and increases in the scale of each enterprise at farm level (Bowler 1986; Gillmor 1987). By the late 1970s, however, it was evident that the policy of encouraging increased output through high price supports could not be sustained indefinitely. Already serious imbalances were occurring in the markets for major agricultural products so that a process of reforming the CAP became inevitable (Commission of the EC 1985a; Philipps 1990). Various measures have been adopted to make European agriculture more market-orientated. These include quota restrictions on output volumes; reduced levels of price support for products in surplus; introduction of market-stabilising mechanisms which limit the quantities of production for which prices are guaranteed; less open-ended and more selective use of the intervention system; an extension of co-responsibility levies; greater emphasis on quality rather than quantity of output; and provision of financial incentives for leaving some land fallow and for afforestation (Commission of the EC 1985b). While most of the measures introduced in the 1980s were of a restrictive nature, the decision to establish in 1980 a common market in sheep meat for the EC provided an opportunity for diversification and expansion. The new regime covering sheep meat was accompanied by high levels of support for producers through ewe premiums which were additional to the headage payments already available to sheep-farmers in disadvantaged areas. These areas include approximately three-quarters of the agricultural land in the Republic.

The declining market for many traditional commodities coupled with the reforms to the CAP have required farmers to reconsider many of the decisions made over the previous decade or more in relation to choice and scale of enterprises. This has led to changes in the structure of production which have altered both the nature and intensity of land use. The extent and distribution of these adjustments are the main focus of this paper. The next section sets out the main trends for the principal enterprises over the past decade. This is followed by a consideration of the effects of restructuring on the scale of operations, which is then related to adjustments in land utilisation at the regional level. The final section considers briefly the implications of the recently agreed reforms to the CAP.

The main sources of data for this analysis are the agricultural census enumerations, the annual sample survey of farms undertaken by the Central Statistics Office, which covers approximately one-quarter of the agricultural area of the Republic of Ireland, and the annual agricultural enumeration undertaken by the Department of Agriculture in Northern Ireland. These surveys provide data on changes in land use and livestock numbers. Additional data relating to the structure of production for individual livestock and crop categories are available from the Farm Structures Surveys which have been undertaken on a biannual basis in the Irish Republic throughout the 1980s. The latest regional-level data from this source are for 1987. While there is a considerable amount of variation within each region, and the accuracy of the data is subject to sampling errors, it is considered to be sufficiently reliable to provide a basis for examining the broad regional adjustments which have occurred.

Overall changes in agricultural production and land use, 1980–90

Agricultural land accounts for 82% and 78% of the total areas of the Republic of Ireland and Northern Ireland respectively. The principal categories used to classify agricultural land use in Ireland are tillage (cereals, root and green crops, plus fruit, horticultural bulbs, flowers and bushes), hay (including grass for silage), pasture, and rough grazing. The relative importance of each of these land uses is summarised in Table 1, which demonstrates clearly the extent to which Irish agriculture is dominated by grassland-based enterprises. Only about 7% of the total area is used for tillage, and even here a very large portion of the output from tillage farms is used to produce winter feeds to supplement fodder conserved from grass.

The principal changes at an aggregate level over the 1980s are summarised in

TABLE 1. Distribution of principal land uses, as percentages of total agricultural area.

	Rep. of Ireland		N. Ireland		Ireland	
	1980	1990	1980	1990	1980	1990
Tillage	9.7	7.4	7.5	6.3	9.4	7.2
Hay	21.3	22.8	24.7	72.5	72.6	75.1
Pasture	51.3	52.7	47.7			
Rough grazing	17.7	17.1	20.1	21.2	18.0	17.7

TABLE 2. Principal changes in agricultural land use and livestock in the Republic of Ireland and Northern Ireland, 1980–90.

	Republic of Ireland		Northern Ireland	
	Absolute change	% change	Absolute change	% change
Total agricultural area (000s ha)	−55.8	−1.0	−14.1	−1.3
Area of arable crops (000s ha)	−138.0	−24.9	−13.5	−16.8
Dairy cows and heifers in calf (000s)	−174.9	−9.7	−8.8	−2.7
Other cows and heifers in calf (000s)	+174.9	+34.6	+21.0	+8.5
Cattle 0 – < 1 year old (000s)	+79.9	+5.0	+14.8	+3.8
Cattle 1 – < 2 years old (000s)	+163.6	+10.5	−16.2	−4.5
Cattle 2 years old and over (000s)	−157.3	−11.1	−24.3	−13.1
Ewes and rams for breeding (000s)	+2746.6	+173.2	+643.2	+115.7
Other sheep (000s)	+2652.5	+155.5	+801.1	+158.8
Horses and ponies (000s)	−14.8	−21.6	−0.9	−12.3

Table 2. While the total area used for agriculture in the island declined by approximately 70,000ha (1.0%), there was a decline of almost 24% (151,500ha) in the area used for tillage crops. Most of the land withdrawn from tillage was used for hay or pasture, but there was also some conversion of former pasture and rough grazing land to forestry and other non-agricultural uses. The net outcome of these changes was that the area of hay, pasture and rough grazing increased by approximately 83,000ha.

The principal trends in the main livestock enterprises which were common to both parts of Ireland were a decline in the number of dairy cows and heifers in calf, and in the numbers of older beef cattle. These declines were more than compensated for by increases in the numbers of cows and heifers in calf retained to produce calves for the beef sector, by the numbers of younger cattle for fattening, and most dramatically by an expansion of the total sheep flock on the island by almost seven million animals. While similar trends were evident in both parts of Ireland, the relative magnitude of most of the adjustments was less in Northern Ireland.

The trends on an annual basis over the 1980s for the major land uses and types of livestock are depicted in Figs 1 and 2. With regard to tillage, apart from a minor increase around 1984, the overall trend has been one of persistent decline. Cereals account for approximately 70% of all tilled land in Northern Ireland and almost 80% in the Irish Republic. The principal cereal crop is feeding-barley, which is used mainly in the manufacture of compound feeds for livestock. However, over the past decade there has been considerable pressure from within the EC to reduce the prices paid for barley and wheat in order to restrict over-production in this sector. Between 1986 and 1990 a combination of cuts in intervention prices and the application of a 3% co-responsibility levy resulted in an overall reduction in cereal prices by 16%. Lower prices resulting in reduced profit margins, adverse weather conditions in the mid-1980s, and a relatively high level of more competitively priced cereal substitutes have all

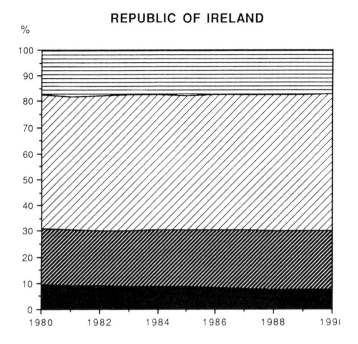

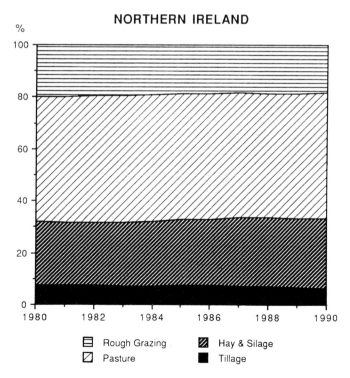

Fig. 1. Percentage distribution of principal agricultural land uses, 1980–90.

REPUBLIC OF IRELAND

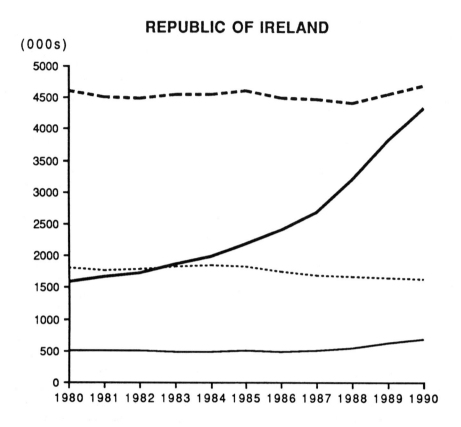

NORTHERN IRELAND

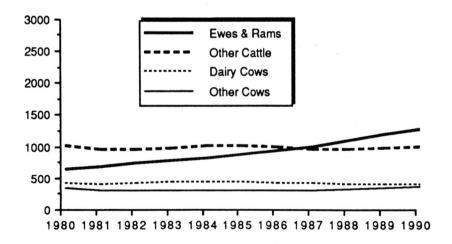

Fig. 2. Principal types of livestock, 1980–90.

combined to reduce the total area of cereals by 116,700ha (26.1%) in the Irish Republic and by 12,200ha (21.2%) in Northern Ireland. Apart from cereals, the only other tilled crops of any significance are sugar-beet in the Republic and potatoes in both parts of the island. The area under sugar-beet has remained constant at about 32,000ha. By contrast the area under potatoes declined significantly by 39% in the Republic of Ireland and 32% in Northern Ireland. This crop, which is not subject to an EC regime, has been under competition from cheaper imports, and has also suffered from poor marketing arrangements.

In the livestock sectors the main trends were the expansion in sheep numbers over the entire decade and the switch from dairying to beef enterprises after 1984 (Fig. 2). The initial stimulus to expansion of sheep-farming came in December 1977 in the form of a bilateral agreement between the governments of the Republic of Ireland and France, under which levy-free access to the French market was granted for 100 tonnes of lamb per week. The securing of this market was extremely important in that it removed the uncertainty and low prices which producers had experienced in previous years when the French market was closed to imports for lengthy periods. Farmers in Northern Ireland benefited indirectly from the new trade agreement, especially in the early years, by exporting live sheep to meat processors in the Irish Republic who were unable to procure sufficient quantities of lamb to satisfy the lucrative French market.

The most important factor, however, in the expansion of sheep numbers was the establishment in 1980 of a common market for sheep meat for the EC. The regulations establishing the new system provided for the removal of barriers to trade in sheep and sheep meat between member states, the limitation of imports from countries outside the Community, and a premium to be paid to producers as compensation when the market prices fell below the Basic (Guaranteed) Price which is fixed annually. The response to the new market situation was particularly rapid in Northern Ireland, where the total number of sheep increased by over 173,000 (16.3%) up to 1982, compared with an increase of 280,000 (8.8%) in the Republic of Ireland.

The final major influence on the choice of farm enterprises was the introduction of the milk quota system throughout the EC in 1984 in order to adjust the imbalance between supply and demand of milk products and at the same time to protect the income of milk producers. The effect of the introduction of quotas on the most profitable grassland-based enterprise was dramatic. In the Republic of Ireland the gain of 167,800 in the number of dairy cows over the decade prior to 1985 was replaced by a loss of 170,000 between 1985 and 1990. The decline in dairying has resulted in a greater emphasis on beef production. The number of cows suckling calves had declined by 30% between 1975 and 1985. However, after 1985 there has been a reversal of this trend, partly in response to rising calf prices and attractive grants especially in the disadvantaged areas, so that by 1990 the beef cow herd had almost returned to its 1975 level. Between June 1990 and 1991 it is estimated that there was a further 10% (67,600) increase in the number of beef cows, while the number of dairy cows declined by a further 13,700 (0.9%).

These macro-scale adjustments which have been outlined indicate that a high level of enterprise substitution has taken place in response to changes in the operation of the CAP. In addition to the substitution response by farmers (which in some cases has involved loss of income), there has also been an emphasis on improving productivity through increased use of artificial fertilisers and compound feeds for livestock. However, the rate of increase in the use of these products slowed down considerably over the 1980s by comparison with the high rates which were part of the intensification process in the 1970s (Harte 1992).

Structural adjustments at farm level

The ability and willingness to respond to the new circumstances varied widely between farms (Harte 1992). The combined effects of reduced profit margins per unit of output, greater emphasis on quality standards by food processors and consumers, and the need for better marketing arrangements in order to improve competitiveness have resulted in greater attention to economies of scale in production at farm level. Consequently, on most farms there has been a tendency towards greater specialisation, which has concentrated the production of each enterprise onto fewer farms. The extent of restructuring is summarised in Table 3. Thus, for example, barley was grown on only 13.6% of farms in 1987 and the average area grown per farm had almost doubled over the previous twelve years. The percentage of farms in the Republic of Ireland with dairy cows declined from 56 to 34 between 1975 and 1987, while the average herd size doubled. Sheep-rearing is the only enterprise for which the number of farms increased over the 1980s.

Adjustments at regional level

The regional framework adopted here is that used by Teagasc, the Agriculture

TABLE 3. Changes in structure of production at farm level.

		Republic of Ireland			Northern Ireland		
		1975	1983	1987	1970	1980	1991
No. of farms (000s)		228.0	221.1	217.0	37.1	30.3	29.4
Avg. agric. utilised (ha)		22.3	22.8	22.7	29.1	36.1	35.8
Wheat:	% of farms with	3.6	2.7	1.8	—	—	2.0
	avg. area grown	5.0	9.3	12.0	—	—	9.8
Barley:	% of farms with	25.0	16.6	13.6	23.2	27.1	17.0
	avg. area grown	4.1	7.3	7.7	5.9	6.3	7.4
Oats:	% of farms with	25.8	9.1	6.1	25.9	6.6	9.8
	avg. area grown	0.6	0.8	1.1	2.0	2.0	3.2
Potatoes:	% of farms with	53.5	32.8	23.9	39.8	25.4	9.4
	avg. area grown	0.3	0.4	0.4	3.1	2.0	3.9
Dairy cows:	% of farms with	55.9	41.4	33.6	39.7	30.6	22.8
	avg. herd size	11.6	18.2	21.8	13.9	29.2	41.0
Other cows:	% of farms with	43.6	32.9	32.7	74.8	67.0	56.5
	avg. herd size	6.7	5.6	6.0	7.9	11.0	15.2
Sheep:	% of farms with	23.5	20.0	21.2	27.9	27.7	38.7
	avg. size of flock	70.2	91.6	108.1	93.0	126.3	226.8

and Food Development Authority, since these are the only units for which comparable data are available (Fig. 3). The locational patterns established by the main enterprises in Irish agriculture in the 1960s and 1970s have been previously examined by Horner *et al.* (1984) and Gillmor (1977). The relative importance of each of the main land uses in each region in 1990 is summarised in Table 4. The most notable contrast is between the importance of tillage in the east and south, and rough grazing in western and northern parts.

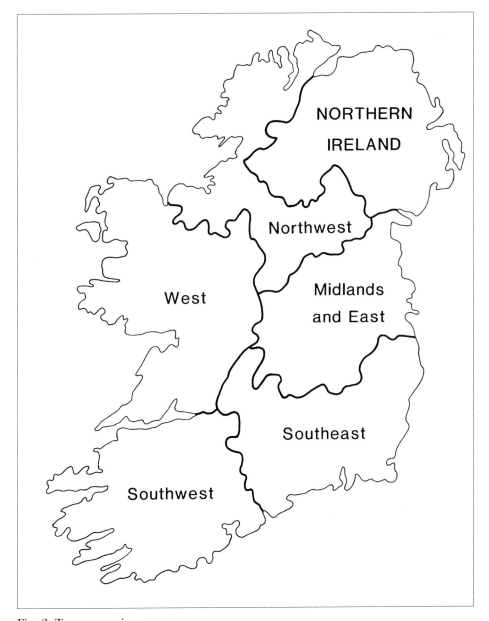

Fig. 3. Teagasc regions.

Tillage

Throughout the 1980s the total tillage area declined by almost one-quarter. The decline was greatest in absolute terms in the Southeast, where in 1980 almost 38% of all the arable farming in the state was concentrated. In relative terms, the most significant decline was in the West, where the total area was almost halved (Table 5). In the Midlands and East and Southeast, where there had been significant increases in the 1970s, there were major reductions in the number of small-scale producers. The contrasts in scale of production between the eastern and western regions resulted in a decline of almost 52% in the number of farms with cereals in the western regions, compared with a decline of 39% for the eastern regions. In Northern Ireland the total number of farms with some cereals declined by 43.6% (from 10,524 to 5654). The decline in the area of cereals cultivated in counties Fermanagh and Tyrone was four times the level of decline in the more ecologically favourable lowlands of County Down.

Dairying

The effects of cutbacks in dairying were most serious on the smaller farms, many of which were no longer able to provide an adequate income either because of restrictions on expansion or because they were unable to invest in the

TABLE 4. Principal land uses by region, 1990 (%).

	Tillage	*Hay and pasture*	*Rough grazing*
Midlands and East	14.9	81.0	4.1
Southeast	13.7	76.0	10.3
Southwest	6.1	72.0	21.9
West	1.3	73.0	25.7
Northwest	2.1	77.7	20.2
Republic of Ireland	7.4	75.5	17.1
Northern Ireland	6.3	72.5	21.2

TABLE 5. Change in total tillage area by region.

	Area (000s ha)			*Percentage change*	
	1970	*1980*	*1990*	*1970–80*	*1980–90*
Midlands and East	143.4	174.6	139.2	+21.7	−20.3
Southeast	179.9	208.9	161.7	+16.2	−22.6
Southwest	113.7	105.6	77.3	−7.1	−26.8
West	52.1	33.3	17.1	−36.0	−48.6
Northwest	43.0	31.4	20.6	−27.0	−34.4
Republic of Ireland	532.1	553.8	415.9	+4.1	−24.9
Northern Ireland	101.6	80.4	64.9	−20.8	−19.3
Ireland	633.7	634.4	480.8	+0.2	−24.2

TABLE 9. Change in number of livestock units by region, 1970–90.

	Absolute change (000s)			Percentage change		
	1970–80	1980–5	1985–90	1970–80	1980–5	1985–90
Midlands and East	32.2	25.6	144.3	3.2	2.5	13.7
Southeast	155.3	33.8	196.6	13.2	2.6	14.5
Southwest	203.7	8.0	73.2	16.9	0.6	5.2
West	92.8	46.3	130.9	8.3	3.8	10.4
Northwest	107.5	19.8	113.0	14.9	2.4	13.3
Republic of Ireland	591.3	133.5	657.9	11.3	2.3	11.1
Northern Ireland	156.8	72.3	133.2	14.0	5.7	9.9
Ireland	748.1	205.8	791.1	11.8	2.9	10.9

TABLE 10. Changes in density of livestock units by region, 1980–90.

	Density*		% change
	1980	1990	1980–90
Midlands and East	1158	1359	17.4
Southeast	1213	1399	15.3
Southwest	1131	1211	7.1
West	917	1062	15.8
Northwest	877	1003	14.4
Republic of Ireland	1056	1201	13.7
Northern Ireland	1236	1440	16.5
Ireland	1085	1239	14.2

* Per 1000ha of land related to livestock.

additional land available for livestock owing to reduced tillage in the east and south, and some losses of land to forestry and other uses, it is evident that the highest rate of expansion has been in the Midlands and East region of the Republic and in Northern Ireland. For the Irish Republic in 1990 the overall density had reached 1.2 LUEs per hectare, and in the most favoured Southeast region a density of 1.4 had been achieved.

The data which have been presented indicate that there has been a high level of enterprise substitution and intensification of land use. The main influences on these trends have been identified as significant adjustments in the policy framework which guides EC agricultural production. Intensification has been facilitated by, among other things, greater use of artificial fertilisers, land drainage programmes, removal of hedgerows and field boundaries, and improved methods of harvesting. Such adjustments in farming practice have placed greater stress on the natural environment (Gillmor 1991–2), which in some areas has become acute (Hickey, this volume).

Further adjustments in agricultural production may be expected in the years ahead. While the role of economies of scale will continue to be of paramount importance, the recently agreed reforms to the CAP have implications for the intensity of each sector of agriculture and for the regional distribution of land uses.

Implications of CAP reform

The package of reform measures agreed for the CAP in May 1992 consists of three interrelated components: (i) supply control restrictions on output from some sectors, especially cereals and dairying; (ii) compensatory payments to offset income losses due to (i); and (iii) payments to encourage diversification, more environmentally friendly agriculture and early retirement of farmers. Clearly it is the supply control measures which have the greatest significance for future land-use patterns. The main proposals in this regard are:

(i)   a cut of 29% in the price of cereals and a minimum of 15% compulsory set-aside for those growers who annually produce more than 92 tonnes;

(ii)   a reduction of 2.5% per annum in the intervention price for butter, plus the possibility of further quota reductions of 1% per annum in milk production in each of the years 1993/4 and 1994/5;

(iii) a reduction of 15% over three years in the intervention price for beef, commencing in 1993/4, as well as a gradual reduction in the quantities of beef taken into normal intervention;

(iv)   an increase in the premiums to encourage expansion of the suckler cow herd and the beef sector, with additional payments on farms with low stocking rates (< 1.4 LU/ha),

(v)   a restriction of the number of ewes eligible for premium payments to 1000 per farm in the disadvantaged areas and 500 elsewhere.

The precise details concerning the implementation of these measures have yet to be worked out, and it will also take some years before their full effects are evident (Fingleton *et al.* 1992; NESC 1992). They will undoubtedly have significant impacts on future agricultural land-use patterns. The price proposal in relation to cereals implies much narrower profit margins, which will probably accelerate the trend of the last two decades to concentrate production onto fewer farms in parts of the midlands and southeast in the Republic of Ireland and the eastern low-lying parts of Northern Ireland. The land released from tillage will probably be used for grazing either cattle or sheep, for which premium payments will be available. The impact of the set-aside proposal is more difficult to assess. While only about 20% of cereal growers are likely to be affected, they account for almost 70% of the total area under cereals. Reidy (1991) estimated that about 34,000ha could be affected by the set-aside proposal. The areas most likely to be affected contain some of the best agricultural soils. It remains to be seen how farmers will adapt to the notion that part of their farms should be considered as a redundant resource, an idea which is alien to the traditional relationship that many farmers have with their land.

The effects of the proposals in relation to dairying can be related to the experience following the introduction of quotas in 1984. Further rationalisation involving a reduction in the numbers of small-scale producers can be expected. This will impact most severely in the more marginal dairying areas, where more farmers may be expected to switch to beef- or sheep-rearing. The proposed payments for beef cattle will probably be of greatest benefit to farmers in the east midlands, where older cattle have traditionally been reared — this sector may expand on land made available as a result of a contraction in cereal-growing. The expansion of the sheep sector will probably continue, though at a reduced rate, over the medium term. Most of the expansion is likely to be in low-lying areas, since the potential for further expansion on many upland and hill areas is now very limited.

Summary and conclusions

Over most of the past two decades decisions concerning the use of agricultural land in Ireland have been taken within the parameters of the CAP. Over the period, the fundamental division between field crops and grassland-based enterprises has been altered very little. However, there have been some significant shifts in emphasis in relation to the use of grassland in response to changes in policy towards agriculture throughout Europe. The main processes affecting agricultural land use have been intensification, specialisation/diversification, and concentration/dispersion. In the 1970s and early 1980s intensification was associated with a significant increase in the use of off-farm inputs and a tendency towards greater specialisation on enterprises for which high prices were guaranteed under the CAP. This in turn resulted in a trend towards greater regional specialisation and concentration of production. From the mid-1980s the trend towards more intensive use of land has continued, even though the rate of increase in inputs such as fertilisers has slowed down considerably. This has been partially in response to declining incomes but it is probably also related to more efficient grazing of grassland following the diversification of cattle and dairy farms towards sheep-rearing. Thus for the most part the intensification process of recent times has been more environmentally friendly, with the exception of some fragile ecosystems which have been damaged by over-grazing by sheep. This unfortunate outcome is related to the unrestricted way in which the ewe premium scheme has operated.

Future land-use patterns will be shaped within the parameters of the reformed CAP. The main thrusts of the reforms are to make agricultural production more responsive to the requirements of food markets, and also to encourage more environmentally sustainable types of farming. As already indicated, some progress has been made in this direction. It has been critically dependent on the opportunities for expansion of. and the levels of assistance made available towards, the sheep sector since 1980. The prospects for further expansion in this direction are dependent on, firstly, the ability of Irish farmers to retain and expand their share of the market for lamb, and, secondly, the availability and magnitude of premium payments. Special attention will be required for those parts of the country where ecological conditions impose limits on the potential for diversification and intensification.

References

Agriculture and Food Policy Review Group 1990 *Agriculture and food policy review.* Dublin. Stationery Office.

Bleasdale, A. J. and Sheehy-Skeffington, M. 1992 Influence of agricultural practices on plant communities in Connemara. In J. Feehan (ed.), *Environment and development in Ireland,* 331–6. Environmental Institute, University College, Dublin.

Bowler, I. R. 1986 Intensification, concentration and specialization in agriculture in the case of the European Community. *Geography* **71**, 14–24.

Boyle, G. E. 1992 National responses to the CAP reform proposals. In *The impact of reform of the Common Agricultural Policy,* 133–68. NESC Report No. 92. Dublin. Stationery Office.

Commission of the EC 1985a *Perspectives for the Common Agricultural Policy.* Brussels.

Commission of the EC 1985b *A future for European agriculture.* Brussels.

Commission of the EC 1990 The Community's agricultural policy on the threshold of the 1990s. *European File,* 1/90. Brussels.

Edwards, C. J. W. 1987 The changing role of sheep production in Northern Ireland. *Irish Geography* **20** (2), 98–100.

Fingleton, W. A., Leavy, A., Heavey, J. F. and Roche, M. 1992 *Impact of the Common Agricultural Policy reforms 1992.* Dublin. Teagasc.

Gardiner, M. J. and Radford, T. 1980 *Soil associations of Ireland and their land use potential.* Dublin. An Foras Talúntais.

Gillmor, D. A. 1977 *Agriculture in the Republic of Ireland.* Budapest. Akadémiai Kiadó.

Gillmor, D. A. 1986 Behavioural studies in agriculture: goals, values and enterprise choice. *Irish Journal of Agricultural Economics and Rural Sociology* **11**, 19–33.

Gillmor, D. A. 1987 Concentration of enterprises and spatial change in the agriculture of the Republic of Ireland. *Transactions of the Institute of British Geographers,* new series, **12**, 204–16.

Gillmor, D. A. 1991–2 Agricultural impacts and the Irish environment. *Geographical Viewpoint* **20**, 5–22.

Harte, L. 1992 Farm adjustment in Ireland under the CAP. In J. Feehan (ed.), *Environment and development in Ireland,* 271–84. Environmental Institute, University College Dublin.

Horner, A. A., Walsh, J. A. and Williams, J. A. 1984 *Agriculture in Ireland: a census atlas.* Department of Geography, University College Dublin.

NESC 1992 *The impact of reform of the Common Agricultural Policy.* Report No. 92. Dublin. Stationery Office.

Philipps, P. W. 1990 *Wheat, Europe and the GATT.* London. Pinter.

Reidy, K. 1991 CAP reform and the cereals production sector. *Farm & Food* (Oct./Dec.), 8–10.

Sheehy, S. J. 1988 Irish agriculture in the nineties. *The Irish Banking Review* (Autumn), 21–32.

Walsh, J. A. 1989 Enterprise substitution in Irish agriculture: sheep production in the 1980s. *Irish Geography* **22** (2), 106–9.

Walsh, J. A. 1991 A regional analysis of enterprise substitution in Irish agriculture in the context of a changing Common Agricultural Policy. *Irish Geography* **24** (1), 10–23.

In: A. Fenton and D.A. Gillmor (eds) 1994 *Rural land use on the Atlantic periphery of Europe: Scotland and Ireland,* 117–30. Dublin. Royal Irish Academy.

FORESTRY AS A LAND USE IN SCOTLAND

Donald G. Mackay

Abstract: The characteristics of Scotland as an environment for forestry are outlined. The development of Scottish forestry is reviewed, concentrating on the major afforestation of the twentieth century and the factors underlying it. A discussion of current issues highlights multi-purpose forestry and indicative forestry strategies.

Introduction

The expansion of forestry is the most dramatic change in Scottish land use in this century, with an approximate trebling of the forest area. The aim of this paper is to examine how the change has come about, and what problems it poses or may still pose — especially where there are parallels between Scotland and Ireland.

Some definitions may be useful at the start. *Forestry* will be taken as the management of woodland, whether natural or planted. *Afforestation* is the creation of a forest by planting bare ground. *Sustainability* (the old term was 'sustention') of a forest is the management of silvicultural operations, e.g. establishment, thinning and harvesting, so that the yield is maintained over a long period of time; or, more technically (Anderson 1967), so that all age classes are present in the forest in sufficient numbers per class to ensure a continuous quota of mature trees.

Next, the approach which is taken in this paper. The subject can be tackled from various angles — the temporal/historical; the spatial/geographical; the modal, i.e. different types of forestry and species; the perspectives of ownership and use; and consideration of forestry objectives. All these are of interest in their own right and by way of comparison with the situation in Ireland; but the scope of this paper is not such as to allow each of them to be developed to the same extent. The temporal/historical approach receives a particularly large share of attention. This is partly because of the long time horizons involved. Not only do trees have the longest rotation of any growing crop, but placing land under trees, especially in the form of plantations, tends to have permanent and even irrevocable consequences for the future of the land concerned.

This study accordingly examines (a) Scotland as an environment for forestry, (b) the development of forestry in Scotland, and (c) current issues and trends.

Scotland as an environment for forestry

Scotland is large enough to have climatic and edaphic contrasts which are

117

significant for forestry – as between east and west, and between north and south. However, it is possible to characterise Scotland as a whole as oceanic-boreal, with cool summers and winters that are not too severe (resulting in a longish growing season), but with winds constituting a problem (Birse 1971). Growth rates for conifers are generally high compared with Scandinavia and even central Europe, but there is great variability over short distances. These features are to some extent shared with Ireland, as is the prevalence of peat and mineral soils. Where Scotland differs from Ireland is in the extent of high ground and steep slopes, and also in the intensity with which the high ground is grazed, especially by deer. These factors limit sharply the areas where trees can grow successfully. However, theoretical land-use capability for forestry is extensive and has been mapped (MLURI 1989).

The altitudinal limit for trees varies according to exposure and latitude, but can be taken as approximately 500m. Below that level the prevailing natural cover is tree cover (Miles 1987), as indeed it is throughout much of the world. The reasons why this cover does not exist at higher levels are considered in the second section of this paper.

The development of forestry in Scotland

As in Ireland, the early story of the Scottish landscape is, first, one of timber exploitation (mainly up to the end of the seventeenth century) and, second, of systematic agricultural development, both arable and pastoral (in the eighteenth and early nineteenth centuries) (Anderson 1967). Both these activities had the effect of greatly depleting and often destroying the natural forest cover, while subsequent developments, especially the establishment of deer 'forests' and grouse moors, have given the natural cover no chance of returning. Recent Scottish forestry, therefore, has been virtually confined to the establishment of new woodland, and it has grown up in the interstices of other land uses. Since forestry has continually had to fight for a place in the sun, it has tended to be single-minded in its aims, as indeed have the other land uses.

The story of Scottish forestry may be started conveniently with Dr Samuel Johnson. During his journey to the Western Islands in 1773 he remarked that from the bank of the Tweed to St Andrews he had not seen a single grown tree, and his companion Boswell, as usual, covered himself with confusion by asserting that there was such a tree a few miles away! Johnson went on to charge the Scots with improvidence, since "to drop a seed into the ground can cost nothing": they did not even have the excuse of the Irish, where it might be argued that an unsettled way of life, and the instability of property, made planting a doubtful investment (Chapman 1924). In this, as in most other things, Johnson was being provocative. In both countries the emergence of the great landed estates had prompted not only much amenity planting, of decorative hardwoods and introduced conifers, but also the beginnings of productive forestry, both on low ground and upland. Anderson (1967) chronicled a thriving timber trade in Scotland, much of it from plantations, from 1790 onwards, and also the management of some deciduous woodland for fuel for industrial purposes.

Commercial planting was, however, spasmodic and had faded out altogether by the mid-eighteenth century, in the face of cheap imports of timber from Canada and elsewhere (Mather 1993). Although the Royal Scottish Forestry

Society can trace its origin to 1854, its membership long continued small and embattled. Anderson (1967) characterised the second half of the nineteenth century as one of "inaction and laissez-faire". Nevertheless there were stirrings even before the turn of the century, and a succession of government committees expressed anxiety about the relative treelessness of both Britain and Ireland, and especially the almost total dependence on imports of timber (Acland 1918).

The Forestry Commission

It took the First World War to translate this anxiety into action. Despite the ruthless exploitation of the remaining forests to meet wartime needs, the supply position became critical. Lloyd George is reported as saying that the war was more nearly lost through shortage of pit props than of food. It was of course too late to put any remedy in train, but at least there was a determination to prevent a recurrence.

The Acland Committee was set up in July 1916 "to consider and report upon the best means of conserving and developing the woodland and forestry resources of the United Kingdom, having regard to the experience gained during the war". It found a state of affairs which it regarded as shocking. The United Kingdom was the most poorly wooded country in Europe except Portugal, Scotland having only 4½% of its area under wood (and Ireland less than 1½%). Average yield per unit area was half of what prevailed in Germany. The Committee recommended the creation of a strategic reserve of 0.7 million hectares to be added to the 1 million or more hectares of existing woodland which needed to be restored and productively managed. Two-thirds of the reserve area would have to be found in Scotland, where there would be particular spin-offs in the provision of rural employment and the opportunity to create a new kind of smallholding.

The agency recommended by the Committee to carry through this policy was an intensely centralised one — a Forestry Commission reporting to no territorial minister, indeed to no minister at all, but only to Parliament. No member of the Committee was more insistent upon this than Lord Lovat, an ardent Scot but one whose concern for forestry exceeded even his nationalism. His reasons for insisting on a single forest authority for the British Isles — set out in a note of reservation because he felt that the Committee's report had not been forceful enough — were, first, to make a definite break with the past (by which he meant the hopeless shilly-shallying of the Scottish Board of Agriculture and other bodies); second, to make sure that forestry policy for the British Isles was not merely decisively planned but decisively put into effect; third, to ensure that forestry was planned and executed for Great Britain as a whole, without undue regard to pressures from any individual country; and fourth, to constitute a body which would be ready in time of war to act along with the military to exploit both state and private forests for the good of the country.

The Acland Report was accepted and implemented promptly and almost to the letter. Thus in 1919 the Forestry Commission (FC) was born, and a decisive move made which has had lasting consequences not only for the scope but also for the distribution and nature of forestry throughout Great Britain. The only concession made to nationalism was through the provision of specific officers, and offices, for Scotland and Ireland. Of course Ireland came out of the FC's remit in 1922, before any significant planting had been done there, and

Northern Ireland never got back in. In 1924 the Labour government (under Ramsay Macdonald) would have hived off Scotland, but Lord Lovat once again persuaded the chancellor of the exchequer to keep the Commission intact (Ryle 1969).

The obverse of the decision to administer forestry on a Great Britain basis was, of course, that policies, e.g. for the acquisition of land or the provision of employment, could not be shaped to meet specifically Scottish conditions or needs. It was even difficult in the early days, as Anderson (1967) complained, to discover how much planting was being done in Scotland from year to year. Opportunism ruled — and indeed it had to, because during a period of national financial crisis the Commission was forced to exist on short commons and against a background of wildly fluctuating government policies, for example a cut in forestry provision coupled with additional funding to counter unemployment (FC 1922). In the same year it had narrowly survived a call for its extinction.

Scottish forestry in the inter-war years

On the part of the private sector there was no significant interest in planting in the 1920s or 1930s. A planting subsidy of £0.80 per hectare was paid from 1921, but it was a time of economic and agricultural depression, and the rate of planting in Scotland remained at about 2000ha per year. These conditions, however, told in favour of the Commission. It was taken for granted by the Acland Committee and subsequently that arable land would not be afforested, and so the FC was restricted to hill pastures and waste ground. But it was able in the circumstances to get hold of hill farms cheaply, and between the wars acquisition in Scotland proceeded at an average rate of about 6000ha per annum (Anderson 1967).

Naturally the FC acquisitions included much unplantable land, but even so the Commission was able to make a virtue of necessity by declaring a series of forest parks, at minimal expense, in some of the most scenic areas of Scotland. This initiative was well received (and counterbalanced the bad image the FC had picked up from some of its planting in the Lake District of England). It marked the beginning of the recreational function of the Commission, and indeed for a time it was bidding to become the National Park Authority for Britain (Sheail 1981).

The Second World War

In 1939 the Commission had a total holding in Scotland of 225,000ha, of which about half was plantable and about a third had been planted (Anderson 1967). Even with the similar area which had been acquired in England and Wales, this represented fairly slow progress towards the Acland goal of 700,000ha. Moreover, all the timber was immature and could make no contribution to the war effort. So the 1914–18 formula was repeated: existing private forests were plundered even before the earlier depredations had been made good. Also, as Anderson (1967) pointed out, fellings tended to concentrate on the most valuable stands of timber and the most vulnerable areas of Scotland.

The FC took the initiative in 1943 by, in effect, rewriting the Acland report for the next fifty years. Its report, *Post-war forest policy*, was forceful, thorough and opportunistic. Instead of a target of 0.7 million ha it boldly went for 2 million ha, justifying it by reference not just to war emergency but to possible future stringency in world supplies. Of this target, 0.8 million ha was to come from the better management through 'dedication' of existing private woodlands, and 1.2 million ha from the afforestation of bare ground — of which the Commission satisfied itself that it could count on an adequate supply, mainly from Scottish hill grazings. Showing considerable foresight, the Commission brought new factors into the forestry case — amenity, recreation, rate of return, and the superiority in terms of rural development of forestry over hill farming. No concession whatever was made on the possibility of combining the administration of forestry with that of any other function. However, devolution of executive business to Committees and Commissioners for England and Wales and for Scotland was proposed, in view of the increased scale of operations which the new target figure implied. This move, which could have been significant for Scottish forestry, was never implemented.

Post-war forestry

The FC was, like the Acland Committee, fortunate in striking at the right time and getting its recommendations substantially accepted. Not only did it secure definite funding for the vital first five years of the planting programme, but the principles it had outlined — notably the formation of large 'forest regions' (e.g. in the Moray Firth and the Scottish–English Border), using a small range of species with a view to large-scale exploitation — went through unchallenged. As Ryle (1969), the historian of the FC put it: what did it matter to the authors of *Post-war forest policy* what proportion of planting was on bare land and what on badly stocked or felled woodland? "The objective would be equally well attained, so far as end-production was concerned, whichever class of land was the more quickly put to efficient use."

It was another fifteen years before another government inquiry blew the whistle on the 'war reserve' argument which had headed the case for expansion in *Post-war forest policy*. The National Resources (Technical) Committee (1957), chaired by Sir Solly Zuckerman, observed that, in a nuclear age, pit-props had little to do with prosecution of the war effort. But by then the post-war planting programme was well launched, with the private sector joining in earnest for the first time. New planting averaged over 20,000ha per annum throughout the 1950s, more than half of it in Scotland, and fully one-third of this was on private estates (see Fig. 1). But the emphasis on home food production which was also gaining momentum meant that forestry was driven further and further 'up the hill' and into the poorer land of the north and west, with a marked restriction of species choice (Anderson 1967; Mather 1993).

It has to be said that this phase of forestry development, which has had a profound effect on the face of Scotland, was characterised by a lack of vision and a tendency to turn its remoter stretches into a tree factory. At a symposium held at the Royal Society of Edinburgh in 1960, Max Nicholson, the guru of the Nature Conservancy, said: "Land use in this country is absolutely hit-or-miss — the basis of the forestry programme will not bear scientific examination". W. A. P. Black, a distinguished research chemist, commented that a system of timber production had been instituted which, by virtue of its long rotation, could not properly be arrested: the problem of disposal of the product had now become of commercial consequence. In reply, James Macdonald, deputy director-general of the FC, accepted the need for a large pulp-mill to absorb the product, but this merely reinforced the need for single-species planting, since manufacturers did

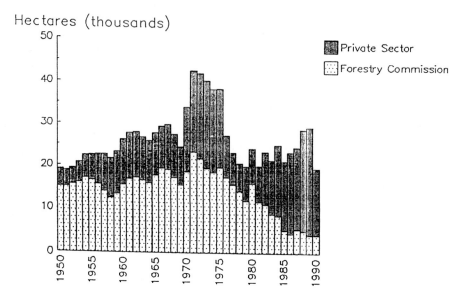

Fig. 1. New forest planting, 1950–90.

not like to vary their chemical formulae, and they needed a cast-iron guarantee of supply. In fact, said Macdonald, a big expansion of Scottish forestry was needed. Why must we accept the current orthodoxy (stated earlier by the director of the Grassland Institute) which restricted "nimble foresters" to the steep slopes and boulder-fields, while "stiff-jointed shepherds" (and sheep) had the run of the smooth and gentle slopes (Scottish Council (Development and Industry) 1961)?

Forestry in Scotland was not to get access to the smooth and gentle slopes for another 25 years. But a pulp-mill did materialise in 1965, at Corpach near Fort William. It failed in the early 1980s, partly for the reason indicated by Macdonald — that its sources of supply were not secure enough (McNicoll *et al.* 1991).

The Treasury epoch

Since the Acland report of 1918 the UK Treasury had cast a baleful eye on forestry and in particular on the FC. By the early 1970s, the technique of cost/benefit analysis had come into vogue, and the Treasury insisted on its application to what it saw as the woefully low rate of return from forestry. A Cabinet Office report of 1971, which proved influential with the government of the day, put a certain value on the recreational and employment potential of forestry, but also called for shorter rotations and less investment at the establishment phase, in order to improve the financial yield. These stipulations are subversive of good forestry practice. Among their effects was a direct squeeze on forestry employment; Mather (1988) quoted a 19% reduction in FC workers between 1981 and 1984. Also there was a move towards clear-felling instead of brashing and thinning — again a measure inimical to resident forestry employment, because the labour required for planting and felling could just as easily be supplied by mobile squads, using the latest techniques in mechanical harvesting, as by resident workers.

The Treasury also found it difficult to leave alone the taxation regime for the private forestry sector, and this had unsettling effects on its performance. The traditional type of planting on private estates collapsed with the replacement in 1975 of estate duty with Capital Transfer Tax. On the other hand, the phenomenon of 'investment' forestry which had begun in the 1960s — the assembly of blocks of farmland by so-called 'forestry groups' seeking to attract capital from private investors — expanded dramatically. In terms of overall planting rates, therefore, the 1970s broke all records, and an average of 40,000ha was planted annually between 1970 and 1975, equally divided between the FC and the private sector (Fig. 1). Such rapid progress was being made, indeed, that in spite of a falling-off in the later 1970s the fulfilment of the 1943 target of 2 million hectares was in sight by the end of the decade, and was reached in 1983, fully ten years ahead of schedule. It was probably no coincidence that the period 1977–80 saw two expansionist forecasts of world timber demand, providing a justification for continuing with production-oriented planting which was accepted by the incoming Conservative government in 1979.

While the decision to maintain an afforestation programme must have been taken in the teeth of Treasury advice, it is not difficult to detect Treasury influence in the government's announcement that, in order to reduce public expenditure, the FC should progressively run down its estate and the emphasis in planting should switch away from the Commission to the private sector. To this latter end, simplified and more generous grant schemes were introduced in 1981 and 1985. These proved successful beyond expectations, so much so that new private planting rose even above the levels reached in the early 1970s (Fig. 2).

The cause, however, did not lie wholly or even mainly in the grant schemes themselves, but in a combination of external circumstances. The emergence of agricultural surpluses meant that the jealous protection of farmland began to weaken, and the price of hill land in particular started to fall. The activity of the forestry groups mentioned above, driven by tax relief factors, reached a new pitch. Environmentalism south of the Border, and the saturation of the best hill areas within Scotland, drove the groups towards the north and west. Some of the resultant planting, especially in Caithness and Sutherland, was by common consent among the most environmentally damaging that Scotland had seen, as well as being questionable from the production point of view.

The public outcry that greeted this type of investment did not go unnoticed in the Treasury, which took its revenge by cutting off all tax relief on new planting in the 1988 Budget. Enhanced grant rates were introduced at the same time. A Farm Woodland Scheme had been introduced a few months earlier, justifying an increased national target of 33,000ha per annum for new planting. This target was now confirmed, with minimum restrictions on planting in Scotland. These palliatives did not, however, prevent a halving in the Scottish rate of afforestation.

The 'new deal' for forestry

Developments within the last year or two have been so sudden and, in a sense, profound as to be difficult to digest. Mather (1993) has speculated that British forestry is going through one of its periodic 'questioning' phases, following which it may be expected to take off again, though perhaps not quite in the same

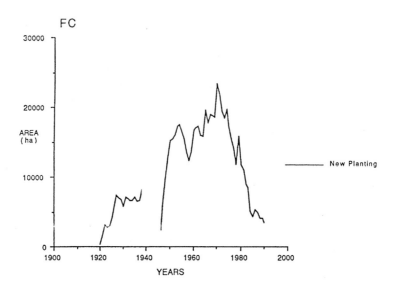

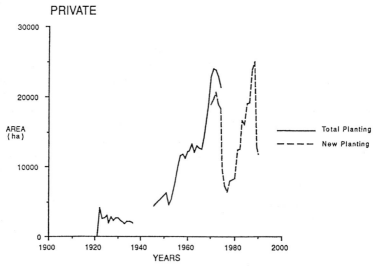

Fig. 2. Forestry Commission and private planting (after Mather 1993).

direction as previously. Certainly forestry expansion is still the watchword, whether in world, EC or UK thinking. But there are signs that, in the UK at least, it will have to proceed within a broader context than ever before.

The major changes so far as British forestry is concerned are as follows.

(a) The FC has been reorganised into two formally distinct arms: a Forestry Enterprise responsible for managing, enlarging and (where so directed) disposing of the existing estate; and a Forestry Authority responsible for standards of forestry, including those of the Enterprise, and also acting as sponsor for the private sector. (A third arm looks after policy and resources for the Commission as a whole, which still functions as a corporate entity.)

(b) Forestry standards now claim to embody fully the concept of sustainability, whether in economic, environmental or social terms. These latter objectives are reflected in a relaunched Woodland Grant Scheme (WGS)(1991), in which maintenance of the forest has been reinstated in 1992 as a grant-aided operation. Broadleaved planting has been encouraged since 1985 by a separate scheme, and though it made no impact on Scotland initially it has since been given greater attractiveness over coniferous planting in grant terms.

(c) 'Protection' of agricultural land is nearly a dead letter. In fact, with minor exceptions, it becomes a virtue to plant trees on productive farmland, and there is even a generous 'better land' supplement to the WGS, in addition to the Farm Woodland Premium Scheme (FWPS) (1992).

(d) Forests are being developed specifically for non-commercial purposes, both for nature conservation and for amenity. The bleak man-made desert which forms much of Central Scotland is the scene of an ambitious multi-purpose woodland project. An additional £950 per hectare is being paid in supplement to WGS for community woodlands.

(e) EC intervention in forestry, long resisted by UK government, is now being actively embraced, in terms both of environmental appraisal and of grant provisions (though with slightly less alacrity than by Irish governments!).

Current issues

The reader, if he or she has had the patience to follow thus far, may well be inclined to ask what is the purpose of the long historical survey through which he or she has been conducted, and what relevance it has to the subject of 'forestry as a land use in Scotland'. Much of the discussion seems to have been about politicking and power struggles affecting Britain as a whole, and very little about the distribution and performance of different varieties of trees in Scotland. It is unfortunately the case, however, that only to a limited degree has forest establishment been practised in Scotland in a way of which foresters themselves can be proud. The places where the greatest concentrations of coniferous forest now occur are not necessarily the places where the landscape or the community benefits, but where land happens to have become available at an advantageous price. The deciding factors have tended to be outside forestry altogether, and often outside Scotland altogether.

At present, Scottish productive forestry extends to 1.1 million ha, i.e. 15% of the land surface. Its distribution is shown in Fig. 3. About 99% of FC forest and 80% of private forest is coniferous. The main varieties are Sitka spruce and lodgepole and Scots pine, with yield classes (in m^3 per ha annually) ranging from 4 to 24, averaging perhaps 10–12 (Table 1). The age profile is shown in Fig. 4. Employment in Scotland is reckoned at 15,000, increasing to perhaps 20,000 by the year 2000. Annual output trends (Fig. 5) show a steady rise (in UK terms) from less than 5 million m^3 in 1980 to a projected 9.5 million m^3 in 2004. All this represents a massive achievement in production terms, and a significant source of employment and economic growth, often in underdeveloped areas. It is the environmental side of the business over which the main issues arise.

Government policy calls for continued new planting at the rate of 33,000ha per annum. Since, by virtue of a government announcement of March 1988, upland afforestation is heavily restricted in England whilst being encouraged in

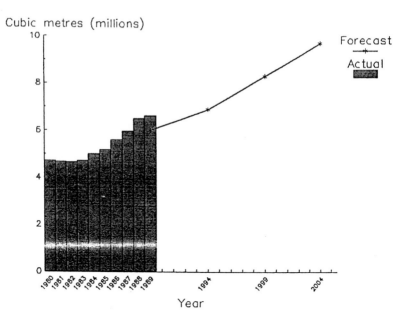

Fig. 5. Trend in timber production for British forests, with projections to 2004.

an attractive option financially, it may not lead to any significant widening of objectives.

Second, the driving force behind forestry is the grant system which, with EC blessing, looks like remaining a permanent feature of the scene. Whatever incentives are offered to incorporate public benefits in private forestry, the fact remains that new forests will result from the individual commercial decisions of a wide variety of private owners, principally motivated by a calculation of the grants on offer. The FC has set out to assist these decisions, and to improve standards of design, by issuing a series of guidelines — on water, landscape, and nature conservation (FC 1988–90) — on compliance with which the grant award may depend. However, a new 'culture' takes time to develop. It is not realistic to expect all at once a dramatic improvement in standards, still less a coherent achievement country-wide of the multiple objectives.

It is for this reason that some have proposed a more radical approach, notably the Scottish Wildlife and Countryside Link, in their discussion paper 'A forest for Scotland' (1992). This calls for a "widely agreed, long term, sustainable forest policy which can be supported by all". It projects the development of three types of forest: an expanded natural forest (to be left unmanaged); an 'extensive' forest, with low input, restocking by natural regeneration, and ownership and management at least partially in the hands of the local community; and an 'intensive' forest, which would be privately owned but managed in accordance with current best practice, that is with thinning and pruning and a good deal of 'biodiversity'.

This is all very well, but the report began to fall apart when it came to considering how such a forest structure could be brought about. A "major review of forestry policy" was called for in the first instance, but the paper was unsure whether the FC was capable of conducting such a review, let alone implementing it. And indeed scepticism is justified, because according to a ministerial statement of 29 May 1981 the FC has the status of a government department and is thus subject to financial and other policy constraints which make it an unlikely sponsor of an open-ended review of this kind.

The likelihood is, therefore, that the attainment of multi-purpose forestry in Scotland will be slower than most people would wish. In the short term, it is more realistic to look forward to a mitigation of the unsatisfactory environmental features of recent practice in afforestation.

Indicative forestry strategies

The concept of indicative strategies, which sounds high-flown and empty of practical significance, has probably as much mileage in it as any of the recent initiatives. This is all the more remarkable because local government, which is the source of these strategies, has no statutory locus as regards forestry.

Consultation with Scottish local authorities over certain types of application for forestry grants has been going on since 1974, though its status is purely advisory so far as the FC's disposal of the application is concerned. Beyond that, for the purpose of the 'structure plans' which regional councils (established in 1975) were obliged to draw up for their areas, the Scottish Office issued National Planning Guidelines from 1977 onwards on various key land-use topics, which naturally included agriculture and forestry. This gave the regional councils an entrée into forestry policy, although because of their restricted locus they were unable to give the topic much substance in their structure plans.

The subject came into sharp focus in 1987 with the controversy over the afforestation of the 'flow country' of Caithness and Sutherland. Following the progressive development of the Irish raised bogs by Bord na Móna, the peat country of the north of Scotland was increasingly prized by conservationists, who saw it as the last European relic of this ecotype and were affronted by the rapid spread of private planting there. The Nature Conservancy Council took up the issue with vigour, and a full-scale confrontation loomed between it and the FC, which supported the cause of the forestry interests. The Scottish Office sought to defuse the issue by inviting the Highland Regional Council (HRC) to chair a working party to examine land-use options in Caithness and Sutherland. Eventually, in 1989, the Council came up with an agreed carve-up of the disputed territory into areas labelled 'Unsuitable' (for commercial forestry) (45% of the whole), 'Undesirable' (20%), 'Possible' (20%) and 'Preferable' (15%) (HRC 1989).

The HRC's approach was hailed as a major advance in land-use planning and was in due course adopted as government policy in Scotland, with the issue to local authorities of Scottish Development Department Circular No. 13/1990, giving guidance on the preparation of Indicative Forestry Strategies. The circular commends a simpler categorisation — 'Sensitive', 'Potential' and 'Preferred' — following a pattern developed by Strathclyde Regional Council. The FC, the forestry industry and conservation interests have generally supported this initiative, which is important because it not only marks out the forestry 'playing field' geographically but indicates within each sector the scale and nature of forestry that is acceptable. This might have been represented in the past as an intolerable restriction on the right of individual owners to do what they like with their own land, but that is not the way it is perceived. It is one of the ironies of political life that the government has recently announced its intention of reorganising local government into a structure of smaller authorities, thus putting at risk the regional dimension within which forestry and other land uses can be effectively planned.

Conclusion

Scotland is still the key area for British forestry, holding the great majority of afforestable and renewable land. However, at a time when forestry throughout most of Europe is steaming ahead, Scotland is standing at a crossroads. Incentives, objectives and constraints have all proliferated at the same time, leaving the industry somewhat bemused. There is no prospect of returning to a *laissez-faire* situation: the danger is rather that with forestry, like agriculture, driven more by grants than ever before, the principles of responsible commercial exploitation and of silviculture may sink even further out of sight. Now, as never before, there is a need for an enlightened and influential forestry authority.

References

Acland, F. 1918 *Reconstruction committee: report of the sub-committee on forestry*. Cmd. 8881. London, Edinburgh and Dublin. HMSO.

Anderson, M. L. 1967 *A history of Scottish forestry* (two volumes) (ed. C. J. Taylor). London. Nelson.

Birse, E. L. 1971 *Assessment of climatic conditions in Scotland: 3. The bio-climatic sub-regions*. Aberdeen. Macaulay Institute for Soil Research.

Chapman, R. W. (ed.) (1924) *Johnson's journey to the western islands of Scotland and Boswell's journal of a tour to the Hebrides*. Oxford University Press.

FC 1922 *Third annual report for year 1921–22*. London. HMSO.

FC 1943 *Post war forest policy*. Cmd. 6447. London. HMSO.

FC 1988–90 *Guidelines: forests and water, forest landscape design, forest nature conservation*. London. HMSO.

FC 1991 *71st annual report and accounts for year 1990–91*. London. HMSO.

HRC 1989 *Summary report and land-use strategy*. Inverness. Highland Regional Council.

Mather, A. S. 1988 Agriculture and forestry. In D. Selman (ed.), *Countryside planning and practice*, 67–87. Stirling University Press.

Mather, A. S. 1993 Afforestation in Britain. In A. S. Mather (ed.), *Afforestation: policies, planning and progress*, 13–33. London. Belhaven.

McNicoll, I., McGregor, P. and Mutch, W. 1991 Development of the British wood processing industries. Paper No. 5 in *Forestry expansion: a study of technical, economic and ecological factors*. Edinburgh. Forestry Commission.

Miles, J. 1987 Effects of man on upland vegetation. In M. Bell and R. G. H. Bunce (eds), *Agriculture and conservation in the hills and uplands*, 7–18. Grange-over-Sands. Institute of Terrestrial Ecology.

MLURI 1989 *Land-use capability for forestry in Scotland*. Maps 1:250,000. London. HMSO.

Ryle, G. 1969 *Forest service: the first 45 years of the Forestry Commission in Great Britain*. Newton Abbot. David and Charles.

Scottish Council (Development and Industry) 1961 *Natural resources in Scotland*. Edinburgh. SC (D & I).

Sheail, J. 1981 *Rural conservation in inter-war Britain*. Oxford University Press.

In: A. Fenton and D.A. Gillmor (eds) 1994 *Rural land use on the Atlantic periphery of Europe: Scotland and Ireland*, 131–42. Dublin. Royal Irish Academy.

FORESTRY AS A LAND USE IN IRELAND

Frank J. Convery and J. Peter Clinch

Abstract: In this paper, the evolution of forestry as a land use in Ireland is discussed, with particular attention to developments in the twentieth century. Since 1981, there has been a radical change in the Republic of Ireland in terms of the volume of planting, the participation of the private sector and the type of sites being planted. Forest planting over the 1982–91 period has been analysed in depth, with particular attention being devoted to private planting. The pattern overall is rapid growth in the area planted, with most of the increase concentrated in the western and northern counties of the Irish Republic. The private sites being planted are highly productive, mainly concentrated on grass/rush sites at relatively low elevations. Sitka spruce is the preferred species, accounting for about 90% of the total planting. Approximately 11% of the total lots grant-aided were referred first for further screening to evaluate environmental suitability. Full-time farmers account for around 65% of the total area planted but investors are concentrated on the larger lots. Environmental issues, including impacts on ecosystems, ecological effects, impacts on aquatic systems and landscape effects, have emerged in recent years as significant factors for debate and analysis, and these are discussed.

Historical context

Forestry as a land use in Ireland involves a paradox: the rate of tree growth is among the fastest in Europe and neighbouring Great Britain is the continent's largest wood importer, yet forests comprise less than 7% of the island's land area. A number of reasons may be posited for such a paucity of trees, but the essence can be summarised as follows: once the indigenous forests were cleared towards the end of the seventeenth century, those who controlled the cleared land rarely felt that the financial and other returns from tree-planting justified the costs. To invest in anything having a pay-off extending fifty and more years into the future, certain conditions are necessary for those who control the land: they must have, or at least think that they have, security of tenure, and expect to retain it indefinitely; they must have resources which can be invested today without prospects of financial return for a generation — such a long view is encouraged if interest rates are low; they must have sufficient land to achieve economies of scale, and have the resources and the authority to limit access; it helps if the other options for using land which yield immediate returns are commercially unattractive to them; they must have a certain faith that the outputs of the forest will be marketable in the distant future.

In Ireland in the past, these conditions were met only for a relatively short time towards the end of the eighteenth century and the first half of the nineteenth, and then only on the part of a small number of landowners. The flowering of estate forestry in this period, although modest in scale, was of great and continuing importance in providing the beginnings of a forestry culture, and in endowing the country with a few woodlands which today are of great aesthetic and environmental significance (Neeson 1991). But most such forest estates did not survive the transfer of the estates from landlord to tenant. The new owners did not have the economies of scale needed to practise effective forestry; they did not have the resources which would allow investment with long-deferred returns; farming became increasingly synonymous with cattle-farming, and livestock farmers everywhere tend to have an antagonistic view of trees as a crop; and finally, forestry was identified as an avocation of the 'landlord class', a group regarded by many as symbolic of an oppressive past rather than as models to be emulated.

At the beginning of the twentieth century, the state emerged as a driving force supporting the restoration of forestry. In Britain and in what was to become Northern Ireland, the ravages and vulnerabilities engendered by the First World War gave the essential impetus. In the rest of Ireland, forestry emerged as a nationalist cause, a *Deus ex machina*, a means of restoring a past which was pillaged, a means of providing the basis for a prosperous future.

The twentieth century

The position of forestry in this century has been well summarised by Gillmor (1993), from whom much of the discussion of this period is drawn. In both parts of Ireland, the state took the lead in restoring forests as a land use. Progress depended on a combination of political will, the associated willingness to provide the necessary resources, and the ability to afforest without 'intruding' on land perceived as being suitable for agriculture. The capacity to expand the boundaries of forest cultivation was enhanced dramatically by the discovery in the 1940s of ploughing techniques which allowed peatlands and podzolised soils to be successfully drained and planted, and the associated development of knowledge of the North American species Sitka spruce and lodgepole pine. With appropriate drainage and fertiliser application, these species could withstand very low-nutrient-status upland sites which were very exposed, and therefore of little value or interest for farming. In the Republic of Ireland, the emergence of this knowledge coincided with very strong political backing for forestry, supported by the rationale that the financial constraint could be relaxed owing to the post-war availability of Marshall Funds. A planting target of 10,000ha annually was set, which, although frequently not attained, endured as the *desideratum*.

State-led forest expansion also characterised Northern Ireland, but at a much smaller scale. Wilcock (1978) noted that throughout the 1970s, annual state planting in Northern Ireland was in the range of 1000–2000ha, whilst private planting was even more modest, not exceeding 250ha annually. Thus, there has been a steady increase in forest area in this century, concentrated mainly on relatively high-elevation and poor-nutrient sites, which were not perceived as being of value for farming, with much of the expansion in both jurisdictions concentrated in the western counties. Farrell (1983) has explained how the constraints on land acquisition were designed to confine afforestation to very low-yielding land, and succeeded in doing so up to the early 1980s. In the years immediately following membership of the European Community (EC) in 1973, the boom in farming resulted in an escalation in the real (net of inflation) price of land, and this put further pressure on the land effectively available to the state. Forestry could not compete with the highly subvented farming alternatives; by 1980, the rate of planting in the Republic of Ireland had fallen to 6000ha.

Throughout this period, the motivation in tree crop establishment was primarily to produce wood, with the subsidiary objective of generating rural employment. But the use of forests as public recreation areas developed also, with a series of Forest Parks being established — commencing in Northern Ireland in 1955, and then developing throughout the 1960s and 1970s in both jurisdictions (Kennedy and McCusker 1983). But the context of Irish forestry was about to change. During the past decade, one could observe alterations in both the magnitude and location of forestry which are unprecedented, and the emergence of environmental considerations as public concerns.

The progress in afforestation in regard to total area planted can be surmised from the data in Table 1.

Forestry in Ireland since 1980 — the (private) phoenix arises

The past decade has seen very striking changes in forestry in the Republic of Ireland; there has been a significant shift in both the volume and the nature of afforestation. Private planting expanded rapidly, and the productivity of the land planted improved. Private forestry heretofore had been confined to a few large estates; new planting on private land was negligible. State planting was confined largely to the relatively unproductive uplands. The key incentive to the private sector has been provided by grants; the tax breaks provided could only be 'captured' at the time of harvest.

During the period while private forestry languished in the Irish Republic, generous tax provisions in the United Kingdom (UK), which allowed the costs of plantation establishment to be written off against current income for tax

TABLE 1. Forest area, Ireland, 1922, 1938 and 1990.

Year	Public forests 000s ha		Private forests 000s ha		Total 000s ha		
	Republic of Ireland	Northern Ireland	Republic of Ireland	Northern Ireland	Republic of Ireland	Northern Ireland	Total
1922	8	2	92	14	100	16	116
1938	14	5	86	15	100	20	120
1990	393	59	81	16	474	75	549

Sources: Graham 1981; Mulloy 1992; reports of Forestry Services; the Northern Ireland data were compiled with the assistance of P. Hunter Blair, Northern Ireland Forest Service.

purposes, resulted in a very dynamic private forestry sector there, concentrated in Scotland. Since the tax write-offs applied to establishment costs but not to land, there was an incentive to buy 'cheap', i.e. relatively unproductive, land, and compensate by applying inputs such as fertiliser and site preparation which could be written off. Under pressure from environmental interests, who argued that the state should not subvent the destruction of key environmentally sensitive upland areas, the generous tax provisions were withdrawn in the late 1980s, and private planting collapsed. As more generous grants have replaced the tax provisions in the UK, there has been a modest recovery in activity. The flowering of private forestry evident in Scotland was not reflected in Northern Ireland.

In the Irish Republic, state planting has been maintained, whereas in the UK, inhibited by the philosophical reservations of the Thatcher era and under threat of privatisation, the involvement of the state has been modest; this lack of enthusiasm is reflected on the ground in Northern Ireland. The combination of anaemic private investment and lack of state activity has resulted in a modest overall performance in terms of planting in Northern Ireland.

Sheehy (1992) has shown that the real price per unit of value added in farming has fallen on average at a rate of 5.5% annually over the 1978–91 period, and this has combined with the introduction of a milk superlevy and quotas to put pressure on returns and limit the attractiveness of farming on land which is judged to be commercially marginal for agriculture. Coincident with the reduction in the returns from farming, the grants available to encourage forest planting were increased sharply in real terms, beginning in the Republic of Ireland in 1981, under provisions of the EC-supported Western Package. The higher grants were continued under the first round of Structural Funding, as embodied in the *Forestry Operational Programme 1989–93*. The programme was in full operation in 1991–2.

It was this combination of 'stick' (falling returns from farming) and 'carrot' (increasing grants for forestry) which resulted in the unprecedented increase in the rate and nature of planting in the 1980s. Some aggregate data from the two jurisdictions are presented in Table 2. The percentage of planting accounted for by the private sector has increased dramatically over the past ten years in Ireland as a whole (Table 3).

Because state planting has increased in the Irish Republic, while private planting has also expanded, there has been a rapid growth in the total area afforested annually. Driven by an aggressive state and private planting programme, the balance of planting activity on the island has shifted towards the Republic over time (Tables 2 and 4). The percentage of the total which is afforested, i.e. that is new planting, is shown in Table 5 for Ireland as a whole.

Private planting in the Republic of Ireland in the 1980s — survey results

For a survey of private forestry planting, the year 1991–2 was taken as being representative of the nature of the land and the type of investor involved. A random sample (approximately 10%) was taken of those who received planting grants in that year, and a variety of information was extracted from the forms completed by the Forestry Inspectors who assess the grant eligibility of the planting proposition in the first instance, and then undertake a grant inspection after planting. These data were compiled and summarised according to a range of criteria, and an interesting and relatively complete picture as to the nature

wood output from existing state plantations is projected to increase from about 1 million m³ in 1990 to close to 5 million in 2010, the latter volume being about 10% of the current production of the largest European producer, Sweden. This volume projection is based on the output to be yielded by the rapid expansion in area afforested which commenced in the 1950s, and which was led by the state. This increase in output will be further augmented as a consequence of the unprecedented rate of afforestation of the past decade, wherein, on the island as a whole, annual planting has increased from about 7000ha in 1982 to about 24,000ha in 1991.

Thus the 1980s have seen important shifts in forestry in the volume of activity, which are attributable mainly to change in the Republic of Ireland. In Northern Ireland the changes are less dramatic, because the reduction in state activity was not compensated for by increased private investment; Northern Ireland's planted area comprises a reducing proportion of total planting on the island, falling from 14.5% in 1982 to 7% in 1991. However, a new and generous grant structure is now in place in Northern Ireland; it is expected that the volume of planting activity will increase as landowners respond to incentives similar to those which have prevailed in the Republic.

The nature of the sites afforested has also changed. Afforestation in the Irish Republic has come 'down the hill'; much new planting on private land is located in the highly productive, relatively sheltered grass/rush lowlands, on land yielding a gross productivity of 20 m³ per hectare per annum. Leitrim, Kerry, Sligo and Tipperary, in that order, are the counties experiencing the greatest 'intensity' of private planting, where intensity is measured on the basis of the percentage of total county area planted privately over the period.

The rapid growth in the forested area has given rise to a number of environmental concerns, focusing on the impacts on ecosystems, especially blanket peatland, the ecological effects on flora and fauna, and the effects on aquatic systems and on landscape. How well have possible environmental negatives been accommodated? The guidelines designed to steer forestry in an environmentally benign direction are in place, and are being implemented, but it is too early to say if further initiatives, or a more regulated approach, will be necessary.

The dominance of Sitka spruce remains unchallenged, comprising 89.6% of the total planting in the sample; the area planted by the private sector under broadleaves is vanishingly small. This latter situation may improve somewhat since the grant differential favouring broadleaves was widened in the new grant provisions which became effective in 1991–2. The environmental referencing system depends for its efficacy on the ability of the forestry inspector to note possible threats posed by planting to the environment, defined in this context to include fisheries, monuments, archaeology, scientific interest and outstanding landscapes. It is important that they have the data and the training necessary to execute this screening system.

References

Farrell, E. P. 1983 Land acquisition for forestry. In J. Blackwell and F. J. Convery (eds), *Promise and performance*, 155–67. Resource and Environmental Policy Centre, University College Dublin.
Farrell, E. P. and Kelly-Quinn, M. 1992 Forestry and the environment. In J. Feehan (ed.),

Environment and development in Ireland, 353–7. Environmental Institute, University College Dublin.

Gardiner, J. J. 1992 Forest production — quantity or quality? In J. Feehan (ed.), *Environment and development in Ireland,* 348–52. Environmental Institute, University College Dublin.

Gillmor, D. A. 1993 Afforestation in the Republic of Ireland. In A. Mather (ed.), *Afforestation: policy, planning and progress,* 34–48. London. Belhaven.

Graham, J. 1981 *Private woodland inventory of Northern Ireland 1975–79.* Belfast. Northern Ireland Forest Service.

Hickie, D. 1990 *Forestry in Ireland: policy and practice.* Dublin. An Taisce.

Kennedy, J. J. and McCusker, P. 1983 State forest amenity policies for a growing urban Irish population. In J. Blackwell and F. J. Convery (eds), *Promise and performance,* 219–28. Resource and Environmental Policy Centre, University College Dublin.

Mulloy, F. 1992 Forestry development — review of existing and prospective EC policies and implementation. In J. Feehan (ed.), *Environment and development in Ireland,* 340–7. Environmental Institute, University College Dublin.

Neeson, E. 1991 *A history of Irish forestry.* Dublin. Lilliput.

Review Group on Forestry 1985 *Report to the Minister for Fisheries and Forestry.* Dublin. Stationery Office.

Sheehy, S. J. 1992 Evaluation of current proposals to reform the Common Agricultural Policy. In J. Feehan (ed.), *Environment and development in Ireland,* 263–70. Environmental Institute, University College Dublin.

Wilcock, D. 1978 Afforestation in Northern Ireland since 1970. *Irish Geography* **11,** 166–71.

In: A. Fenton and D.A. Gillmor (eds) 1994 *Rural land use on the Atlantic periphery of Europe: Scotland and Ireland*, 143–51. Dublin. Royal Irish Academy.

RECREATION AS A LAND USE IN SCOTLAND

Jean Balfour

Abstract: The paper briefly outlines the historical background of the developing appreciation of the Scottish countryside by visitors, and of changes in land use up to the present day. The new interest in the environment and countryside recreation in the 1960s is described. This culminated in the setting up of the Countryside Commission for Scotland, with new powers to support conservation, visitor management and facilities and the development of a ranger service. The implications of the report *A parks system for Scotland* and the attitude of the government towards the planning and management of the Cairngorms and Loch Lomond are highlighted. Increasing demands on the countryside for recreation and the conflicts with other land uses are discussed. Proposals for new initiatives in recreation management are outlined, along with the need for more resources and greater government commitment. Finally, a Celtic initiative is suggested.

Setting the scene

The qualities of the Scottish landscape have long been recognised by visitors to the countryside. Perceptions, however, have changed over the centuries. This is a reflection, perhaps, of the heavy task of wresting crops from often wet and sour land, and the back-breaking work of removing stones. Before the eighteenth-century romantics, therefore, the high hills and uplands were looked on as desolate, and, earlier still, the pine forests as a harbour for robbers and wolves. Improved land was preferred, with its more ordered landscape and productive capacity.

Though Scotland is not an island, access remained difficult. It was General Wade's new roads and the 'Age of the Enlightenment' in the eighteenth century which created a new perception of the wild hills and lochs. This is epitomised in Sir Walter Scott's writings, where *The Lady of the Lake* can be regarded as the first tourist promotion for the Trossachs, which have been much visited ever since.

Today, the whole variety of the Scottish countryside is valued — the hills, the glens and lochs, the woods and forests, the farms and rivers, and the often-spectacular coastline.

Land improvements in the eighteenth century, the Highland clearances following the introduction of pastoral sheep-farming, and the farming depression after 1870 with the subsequent creation of the Highland deer forests were all to

143

have long-term implications. At the same time, forest cover was reduced to an all-time low. The development of the new forests and the increased development and prosperity of agriculture until the mid-1980s are discussed in other papers in this volume. Today both foresters and conservationists would point to the red deer explosion (for example, the doubling of deer numbers between the 1960s and 1980s in areas of the Cairngorms) as a matter of major concern, while farmers and landowners are concerned about the Common Agricultural Policy (CAP) and the increase in numbers of people walking over the countryside.

The landholding structure in Scotland and Ireland is very different. Scotland, because of its generally poor land, has large farms and estates. Ireland, mainly owing to restructuring in the last century, has a multitude of small owner-occupiers and comparatively little woodland. Forests in Scotland, mainly those owned by the Forestry Commission, provide a large and very important recreation resource, including Forest Parks.

In Scotland, recreational facilities are provided mainly (but not exclusively) by local authorities, the Forestry Commission (FC) and voluntary bodies such as the National Trust for Scotland (NTS). Most, though not all, of the 'open hill' country outwith forests is privately owned. About 10% of the north and west, including the islands, is subject to crofting tenure.

Scotland has a plethora of government departments and agencies involved in countryside recreation. However, the Countryside Commission for Scotland (CCS) had an overall remit to encourage both conservation and public enjoyment while having regard to the socio-economic aspects of the countryside. The recent amalgamation of the CCS with the Nature Conservancy Council for Scotland to form Scottish Natural Heritage (SNH) has brought nature conservation specifically into the countryside remit. This has strengthened the function though not necessarily the funding. This arrangement has similarities to the situation in Northern Ireland, where there has been a combined (though previously inadequately funded) countryside service since the 1960s, but differs from the arrangements in the Republic of Ireland where there is no overall countryside agency.

Background

Scotland's population, unlike that of Ireland, is spread unevenly over the country. About 85% are concentrated in the urbanised central belt, and the rural population spreads, sometimes very thinly, over the rest of the country. Scotland therefore combines urbanisation and remote countryside in a manner not found elsewhere in Europe. This pattern has affected, and still affects, the attitudes and opportunities of Scots in terms of recreation and conservation. For other parts of the UK and beyond, the Scottish countryside has been a goal for a range of holiday-makers for some time.

Though there has been interest in access to the hills since the mid-nineteenth century at least, the numbers who walked and explored them were small. Roads were poor and trains limited, while crossing by ferries could be hazardous. In the central belt, many of the towns and cities reflected the upsurge of the industrial revolution, with conditions and working hours which did not recognise leisure and with only very occasional holidays, such as those by train to Balloch on Loch Lomond, or 'Doon the Water' on the Clyde.

The big changes came after the Second World War, when longer holidays and greater mobility created opportunities to travel more widely. Thus, areas which had been relatively unvisited became more widely known, and the need for visitor facilities such as car parks and public toilets came to be recognised. Since many of these visitors came either from the towns or from outwith Scotland, the local authorities, which were called on to make the necessary provision, felt that burdens created by other ratepayers were unnecessarily great. Nor, of course, were visitors only Scots, since increasing numbers came from other parts of the UK and from mainland Europe.

For many of the Scottish population who lived in the urban central belt, countryside recreation was needed nearer at home, if it were to be enjoyed on a more regular basis. Areas of countryside were needed which could be reached without the use of cars.

A new look

The 1960s marked a period of increased environmental concern, linked with increasing visitor pressure on the countryside. It was these twin pressures of care of the countryside and the need to provide for countryside recreation that led to the passing of the Countryside Acts and the setting up in the UK of the two Countryside Commissions and the Countryside Branch of the Department of the Environment, Northern Ireland.

All that seems a long time ago, but it would be wrong to overlook the growth of environmental and recreational awareness that took place at that time. In a more caring society, the opportunity for ordinary folk to be able to enjoy the countryside was seen as an important social objective, linked with better care of the countryside. Public money was used in the necessary provision of interpretative as well as practical countryside facilities. This support was, in Scotland, channelled through the CCS, which, in 1968, started its developing partnership with local authorities, voluntary bodies and other landowners and occupiers.

More countryside recreation opportunities for the urban dweller of limited resources and conservation of the countryside, cooperating with other land use, were the two strands running through CCS policy and strategy through the 1970s and 1980s.

Crucial in this approach was the creation of a ranger service, with a management role in regard to visitors to the countryside. This is to guide visitors, in order to protect sensitive countryside and other land uses, but also to increase visitors' awareness and understanding, so that this in turn leads to greater enjoyment and appreciation. Rangers, particularly in Country Parks, can link into the education system by supporting school groups and teachers. They can defuse land access problems with tact and discretion. At present they are able to operate in country parks, designated footpaths and areas covered by access and management agreements. The Scottish Ranger Service has been traditionally (if twenty-odd years can create a tradition) employed by a variety of employers — local government, voluntary bodies and landowners — but 'united' by national training, national advice and national grant aid, provided by the CCS, which is now the responsibility of SNH.

The development of Country Parks, mainly in and around the central belt, was seen as a key in the provision of easily available countryside recreation for the urban dweller (see Davidson, this volume, Fig. 3). There were 36 of these (covering 6000ha) by the late 1980s. They sought to provide, in a countryside setting, recreation for local inhabitants. This varied from park to park, depending on the local authority, its land, resources and initiative, with varying emphasis on education, wildlife interest, sport and just walking.

Country Parks are owned by local authorities. They have been financed and developed in partnership with the CCS, which provided a share of the investment, skills and expertise, and support for a ranger service. The provision of Country Parks was also important in developing the awareness of local authorities of the importance of countryside recreation at a time when it was for them a new idea. Of course, local authorities were not alone. The FC and the NTS, for example, developed new initiatives, the latter with support from the CCS, while a few private estates, aided by the countryside grant, went into the recreation business.

The countryside of hills and uplands is very different from that of Country Parks, which are dedicated primarily to recreation. The hills are used for grazing and for growing trees and they are used by plants and animals, while the more distant mountains provide qualities of remoteness. The most difficult conflicts arise in the hills and uplands between recreation and other land uses and between competing kinds of recreation. An example of this was demonstrated in two public enquiries over proposals for new skiing developments in the Cairngorms.

The report *A parks system for Scotland* (CCS 1974) recognised the need to provide a spectrum of recreational provision and experience related to conservation and other land use. Designated areas were seen as part of this approach. Today these include Country Parks, whose primary land use is recreation. Regional Parks (the responsibility of Regional Authorities and four in number), which consist of large areas of pleasant (and in the case of Loch Lomond, outstanding) countryside, in both public and private ownership, are subject to significant visitor pressure (see Davidson, this volume, Fig. 3). Within Regional Parks intensive recreation takes place in some parts and extensive recreation along with traditional land use such as farming and forestry in others. The Regional Authority provides overall coordination and a ranger service, as well as recreational facilities. The report also proposed a category of special parks to provide a planning and management authority for areas of outstanding landscape/conservation interest which were under significant visitor pressure, primarily the Cairngorms and Loch Lomond.

The provision of designated areas is complemented by footpaths, particularly in the countryside near the towns but less so in the hills and uplands where, generally, there is open access. The provision of long-distance routes — the West Highland Way, the Southern Upland Way and, more recently, the Great Glen Way — have been generally welcomed, provided they are not seen to encroach on areas of remote countryside.

This general approach — the recreation spectrum, related to conservation and other land use — is still valid today. It recognises, too, the place of sensitive areas such as nature reserves and prime landscape areas.

Designated areas, footpath provision and the provision for general access are not always enough. Positive steps may be needed in a given area to improve landscape and to deal with particular land and visitor problems, whether in the

Edinburgh Green Belt or in the sandy beaches of the northwest such as Achmelvich. These local problems or initiatives need a coordination of interest and available support, an understanding catalyst in the local community, and a discreet leadership which can help things to happen. People to carry out this task have been described as Project Officers, who might stay for two to five years, long enough for the initiatives to have become a part of local activity and commitment. This mechanism has been used quite widely and successfully over the last twenty years in Scotland, but also, for example, in the Mourne Mountains in Northern Ireland.

Carrying on

Better education, more leisure and more mobility, at least for some, mean increasing demands for footpaths, particularly within easy reach of towns, and from more people to walk the hills. While hill-walkers and climbers still constitute only 10% of the walking public (CCS 1991–2; SNH 1992), the imprints of their feet, even on the more remote hills of the far northwest, suggest that they are increasing and that, in due course, they may threaten the very qualities they come to seek. Water, that other resource, often an integral part of the land, is also being widely used, with an increase particularly in the use of powerboats and by water-skiers (CCS 1991–2).

There are arguments about access, demands for the right to roam, difficulties in access to water, and questions about how to manage water for the competing interests of angling, powerboats, sailing or canoeing (Wilkinson and Waterton 1991). There are conflicts with farming, sporting, nature conservation and even forestry, from walkers, bicyclists and those who use non-traditional vehicles. There is concern about conserving the heritage of Scotland, its wildlife and vegetation, its remote areas, and about erosion. The subarctic hills are more fragile and therefore less able to take the pressure of many feet than areas south of the Scottish Border and in Ireland or in alpine mainland Europe. There is concern, too, about jobs and about the potential for job creation through tourism. There is increasing uncertainty about the future for agriculture. More people are interested in the development of native woodlands, though the future for timber producing forests remains unclear.

Land use for conservation cannot be separated from recreational use, since both have an impact on each other and on the countryside which they use.

The talk, then, has become louder, but what about the solutions, or at least steps that could help and improve the situation at a time when there is still scope to do so?

What next?

The late Anthony Crossland said that prosperity in the countryside was a necessary basis on which to build conservation — and, it could be added, recreation. The current uncertainties over the CAP, the drop in farm incomes in the UK and the lack of clear directions in forestry are unhelpful in creating a sound and prosperous basis for countryside conservation and recreation.

How, then, should those concerned build further on the experience and achievement of the past? What now are the main issues which need to be faced?

1. Access to the wider countryside, particularly the hills and uplands.
2. The creation and maintenance of footpaths/tracks/designated ways, in the countryside, around towns and in the hills and uplands.
3. The development of Country Parks — developing more active pursuits for young people.
4. The use and regulation of inland water.
5. The planning and management of areas of prime conservation (nature conservation and landscape) interest which are under significant visitor pressure.
6. The Ranger Service.
7. Information and monitoring.

Access

Contrary to some views, there is no right of access to unenclosed land in Scotland. Under civil law, people can be asked to leave any private land, and camping without permission is a criminal offence. However, in practice, outwith the stalking season, people have generally been free to walk the hills and open country unimpeded. Nevertheless, there is an increasing demand for 'a right to roam' and a wish (perhaps unwise!) to have the present position formalised. Not unexpectedly, any such formalisation is opposed by landowners and occupiers.

There may, however, be arguments for modifying the law for individuals (not groups) for walking only, without dogs, and under quite specific conditions, covering, for example, erosion, damage to property, stock, crops, timber, sporting facilities and conservation requirements. Such arrangements could quite specifically include the non-use of mountain bikes and/or all-terrain vehicles.

Such an arrangement would have to give the owner/occupier the right to demand an access agreement, with the necessary ranger service provided by the local authority and funded by SNH, where problems for land management or conservation could be demonstrated. There would be significant financial implications.

Footpaths\ /tracks/designated ways

Footpath networks need to be further developed around towns, which is where there is a growing demand (CCS 1990). In the hills there still exist today some of the excellent well-constructed and unobtrusive hill paths created by nineteenth-century landowners. Many are in need of repair, and there is a good argument for the provision of some new ones, carefully selected, in hills and uplands which are well visited and which are not in remote countryside.

It has been estimated that some £2 million per year is required for proper footpath maintenance alone. So far, resources have been quite inadequate to tackle this properly. The role of footpaths is a key element in any countryside strategy which takes account of forestry, farming, conservation, public enjoyment and visitor management (CCS 1991–2)

The provision, however, of too many tracks, often for sporting access, is viewed with some suspicion. There are some planning controls but these exist mainly in National Scenic Areas and, it could be argued, should be applied to all tracks in open countryside.

The development of Country Parks

Since 75% of Scots go walking with children or the dog, walking has been an important part of countryside activity (CCS 1990). In general, however, there is scope for Country Parks to develop more active pursuits for teenagers and young people This should be seen as part of a social provision, which could link into urban activities on the one hand and into projects in areas of not-too-distant countryside on the other. Management agreements could be made with landowners for special activities.

The use and regulation of inland water

The conflicts between powerboating and quiet use of water can be tackled by zoning and by management. It is possible for local authorities to make access agreements, but these are very time-consuming as there are usually a number of owners involved — as experience on Loch Ken and Loch Tay has demonstrated. Furthermore, there are problems of access to water which may mean over-concentrations of people at certain places on a loch side, which may not be ideal. New arrangements and special incentives are required to achieve water zoning and coordination of use of designated lochs. This is likely to require stronger legislation if effective management is to take place.

The planning and management of areas of prime conservation

The 1981 Countryside (Scotland) Act, which provided enabling legislation for Regional Parks, failed to include provision for special parks. This now means that, unlike most countries in Europe (and beyond), Scotland has no framework for the planning and management of areas of outstanding conservation and heritage interest which are also under high visitor pressure. Successive governments over more than 40 years have steadily refused to concede this need in Scotland whilst overtly encouraging and supporting such a framework in England and Wales. The state of affairs in the Cairngorms in particular and in Loch Lomond is becoming an international scandal, while at the same time the government seeks to promote the Cairngorms as a World Heritage Site.

The proposal, whether or not it is called a national park, was put forward yet again in the Popular Mountain Areas report of the CCS and, on the basis of a poll carried out by the Scottish Office, is supported by 90% of the Scottish population. Yet the government continues to ignore the need for a proper planning and management framework for these areas. This therefore remains a significant gap in a strategy for the countryside.

The Ranger Service

Earlier in this paper, reference was made to the particular structure of Scotland's Ranger Service, which collectively now employs 160 rangers — a provision which initially received 75% grant aid from the CCS, though this has been reduced in recent years. The provision of an effective and expanded ranger service is central to developing better visitor management and countryside conservation.

The concept of different employers who share national training, national advice, a national ethos and national logo is still crucial. From 1 April 1993 there is potentially a new element in the Ranger Service, since SNH, unlike the CCS, can and does employ its own ranger (warden) service, and this should be seen as an addition to the Scottish Ranger Service. Some of the SNH rangers may be concerned primarily with nature reserve management, but here too there is often a visitor element, and the need to promote better understanding. Rangers may also need to become more involved in land-use issues locally as they affect visitors. Experience in the Northern Ireland warden service suggests that rangers can become well able to handle countryside/nature conservation interests along with land management issues and visitor management.

Information and monitoring

What do people want, what is happening in terms of the patterns of recreational use in the countryside? Problem areas, new initiatives, pressure on the countryside, effects on conservation, relationships with other land uses and remote area experience: all these require monitoring and the regular collection of information to provide the necessary knowledge on which to base strategic planning, management and resource requirements.

The way ahead

Recreational use of the countryside will always be bound up with conservation and with other uses of the countryside. There needs to be a spectrum of recreational use, extending from the intense, in for example Country Parks, to the very light, in sensitive and remote areas. In between there should be a wide range of informal recreation opportunities with proper safeguards on farms and in forests. The scale of acceptable numbers will vary along this spectrum, but arguments that suggest that remote areas are not important because few people go there are not acceptable. Scotland has no wilderness areas as, for example, in Canada or Greenland, but because of their subarctic/oceanic character its northern hills and uplands carry very small populations. They therefore have qualities of remoteness that are increasingly rare in the European context, and are of immense value, since they provide a special experience of the natural world, which people discard at their peril.

Sustainable recreation cannot, therefore, be looked at in isolation and should be seen as a part of overall countryside planning and strategy. How can this be done? Structure plans and local plans do not always deal knowledgeably and effectively with countryside matters, and though built recreational facilities are the subject of development control, rural land use (forestry apart) is not. However, a better way to safeguard the use of the countryside needs to be developed if the new word 'sustainability', currently bandied about, is to mean anything!

Consideration should be given to preparing 'a framework of intent' for the countryside, identifying, for example, land capability (agriculture and forestry), Less Favoured Areas, sensitive areas and currently designated areas, and including a broad view of preferred land use and activities, including recreation. This could provide a strategic framework for the preparation of indicative regional countryside strategies, similar in concept to the indicative forestry strategies which have been pioneered by Strathclyde Regional Council and supported by the Scottish Office. All these steps would require consultation and the working together of planning authorities, relevant government agencies and the Scottish Office.

The management of visitors to, and conservation of, the countryside and other land uses will depend on the development of partnerships on the ground between SNH, local authorities, landowners and voluntary bodies. It depends crucially on the investment of real resources, in people and facilities, in footpaths, in developing Country Parks, in an expanded ranger service, in realistic access and management agreements and 'top-up schemes', in conservation projects and in countryside rehabilitation.

Finally, sustainable recreation will depend on the political will of government which has, as yet, to develop a real commitment to conservation and the countryside.

A Celtic initiative

What, then, can the Scottish experience offer to Ireland? Links with Northern Ireland are quite well established and there is already some sharing of common experience and legislation, but this is different of course in the Republic of Ireland. Two aspects could perhaps be considered.

First, a sound recreation strategy must be based on conservation. In this context the development of indicative countryside regional strategies could prove a mechanism worth investigating.

Second, access to the countryside in Ireland is likely to increase as it has done in Scotland. The development of some kind of ranger service, and arrangements for management and access agreements could be useful tools both for visitors and the land-users.

Given the official will and support, why not a Scottish/Irish (North and South) countryside working group to look together at these aspects and learn from each other's experience?

Acknowledgements

The author wishes to thank M. Payne, J. R. Turner and J. Mackay, formerly of the CCS and now with SNH.

References

CCS 1974 *A parks system for Scotland*. Edinburgh. Countryside Commission for Scotland.

CCS 1990 *Day trips to Scotland's countryside, 1987 to 1989*. Edinburgh. Countryside Commission for Scotland.

CCS 1991 *The mountain areas of Scotland: conservation management*. Edinburgh. Countryside Commission for Scotland.

CCS 1991–2 Unpublished papers and reports on recreation. Countryside Commission for Scotland.

SNH 1992 Enjoying the outdoors: a consultation paper on access to the countryside for its enjoyment and understanding. Scottish Natural Heritage.

Wilkinson, D. and Waterton, J. 1991 *Public attitudes to the environment in Scotland*. Edinburgh. Central Research Unit, Scottish Office.

In: A. Fenton and D.A. Gillmor (eds) 1994 *Rural land use on the Atlantic periphery of Europe: Scotland and Ireland*, 153–71. Dublin. Royal Irish Academy.

RECREATION AS A LAND USE IN IRELAND

H. John Pollard

Abstract: This paper reviews five aspects of rural recreation in Ireland: recreational pursuits and their location, the provision of access to recreational land, the relationship between recreation and the environment in the context of the coast (Ireland's principal recreational resource), the management of that coastal resource, and future demands upon recreational land. Ireland is fortunate in that limited industrial and demographic pressures have permitted the retention of many unspoiled recreational landscapes to which access has never been perceived as a significant problem by users and planners alike. This does not preclude localised environmental problems, particularly through recreational pressures on the more intensely used stretches of coastline in close proximity to Dublin and Belfast. Management strategies have largely coped with present levels of use, but demands from many private and public sector sources are increasing inexorably, requiring close attention to be paid to the safeguarding of recreational resources into the twenty-first century.

Introduction

To deal with recreation and recreational land use in Ireland in a few thousand words is a major exercise in selectivity. To some degree or other all people, both residents and visitors, are users of recreational facilities. Moreover, most of the Irish countryside serves a recreational function, whether people gaze upon it passively on a Sunday afternoon drive, or whether they favour particular parts of it when engaging in any of the more active sporting or walking pursuits. Nor can recreation be seen in isolation from other activities and demands, for it clearly impinges on other land uses, both commercial in the case of agriculture and forestry, and conservational in terms of the need to protect the landscape for a variety of reasons, including its recreational use in future generations. Finally, there is the added dimension in the Irish case of the political division of the island, which might be expected to affect the organisational structures and management of any land set aside for recreational purposes.

That being said, emphasis within this broad and complex topic is placed upon certain themes, namely:

1. recreational demand as indicated by recreational activities and the areas used for recreational pursuits;
2. the provision of access to recreational land;
3. the interaction between recreation and the environment;
4. the management of recreational resources, particularly the safeguarding of recreational land;
5. future demands for recreational land.

Recreational activities and recreational land

Acknowledging recreational demand is a simple matter; measuring it accurately is much less straightforward. Effective demand is essentially the actual participation in recreational pursuits which can be undertaken by local residents on a day excursion, or by tourists coming both from within Ireland and from beyond its shores. Of those sources, regular surveys of tourist activities are much more readily available than those of day-trippers. While it might be argued that there is a fair degree of overlap between the two, as relaxation is a common objective, there are differences between the two groups, just as there are between and within foreign and domestic tourist groups, each in search of their particular holiday experience. Thus, there is need to use some discretion in extrapolating conclusions from one group of users to another.

The principal sources of Irish tourism studies are those undertaken by the two tourist boards — the Northern Ireland Tourist Board (NITB) and Bord Fáilte Éireann. Visitors to both territories are continuously monitored and the results published annually. In Northern Ireland a leisure day-trip survey was also undertaken in 1986–7 (Coopers and Lybrand 1987) and another in 1991 (NITB *et al.* 1992), although the results of the latter at the time of writing are still in provisional form.

Tourism

Among the tourist data collected, information is obtained regarding the principal activities undertaken during the visit of the almost 1.5 million holiday-makers (excluding those visiting friends and relatives) who came to the Republic in 1990. These are set out in Table 1. Unfortunately there are no equivalent data for Northern Ireland for the same date, although, given the similarity of tourist resources, there is no reason to believe that activity patterns would vary greatly.

Both natural and cultural attractions figure strongly, both as important activities while in Ireland and as vital influences on the choice of Ireland as a holiday destination. The heritage of the island is clearly a major 'pull' factor, influencing the choice of substantial numbers of visitors to historic sites and stately homes and gardens, but high proportions of tourists concerned with active pursuits come specifically and often exclusively to enjoy the opportunities for angling, golf and cruising. Many of the passive pursuits are more generally concerned with touring and enjoying the aesthetic appeal of the Irish landscape, which becomes more apparent when the geographical pattern of tourism is examined.

Some data are available on the spatial distribution of tourism and, to a lesser extent, tourist activities, from which can be gained an impression of the geographical distribution of recreational resources (Fig. 1). Unfortunately, tourist data for Northern Ireland and the Irish Republic are not presented on a

TABLE 1. Tourist activities in the Republic of Ireland, 1989–90 (after Bord Fáilte 1991a).

	Involvement in	Influenced choice
Active pursuits		
Hiking/walking	188,000 (11%)	38,000 (2%)
Fishing	150,000 (9%)	91,000 (5%)
Golf	132,000 (8%)	67,000 (3%)
Cycling	92,000 (5%)	30,000 (2%)
Equestrian	45,000 (3%)	18,000 (1%)
Cabin cruising*	51,000 (3%)	31,000 (2%)
Passive pursuits		
Historic site visits	617,000 (36%)	97,000 (6%)
Stately home visits	318,000 (18%)	39,000 (2%)
Garden visits	337,000 (20%)	39,000 (2%)
Genealogy	95,000 (6%)	53,000 (3%)
Pilgrimages	81,000 (5%)	10,000 (1%)
Horse-racing	54,000 (3%)	11,000 (1%)
Language study	49,000 (3%)	20,000 (1%)
Bird-watching	42,000 (2%)	6,000 (~)
Dog-racing	20,000 (1%)	3,000 (~)

* Includes 'Other water-based activities'.
~ Less than 0.5%.

compatible basis. Figures for Northern Ireland include both home holiday-makers and visitors, and are distributed according to the administrative subdivision of the territory into district council areas. Figures for the Republic refer to overseas visitors only, and are allocated on a county basis. A further complication is that the pattern of holiday-making is not immediately apparent as the data relate to tourism of all types, including that for social and business reasons. Thus the pattern inevitably displays a residential population effect, manifested in terms of 'the larger the population, the more friends and relatives would be likely to visit', as well as the pull of the main commercial, industrial and administrative centres (Pollard 1989). Such effects support the dominance of Dublin in the Republic of Ireland and raise Belfast to the number one position in Northern Ireland.

The data can, however, be reworked to take account of the population effect, so providing a clearer impression of the attractiveness of different parts of the island for holiday-making pursuits. Figure 2 is based upon such revised data, the map showing *tourism quotients* which describe the distribution of tourists in comparison with that of population. All counties or district council areas with quotients in excess of 1.0 have a level of tourism activity in excess of that expected on the basis of population alone. A clear picture of high- and low-demand areas can be seen, reflecting the distribution of Ireland's main attractions (Fig. 3). Compared with Fig. 1, Dublin and Belfast are severely downgraded. In the Republic of Ireland, a strong western bias generally shows through as tourists congregate in the scenic mountain and coastal belt, ignoring

156

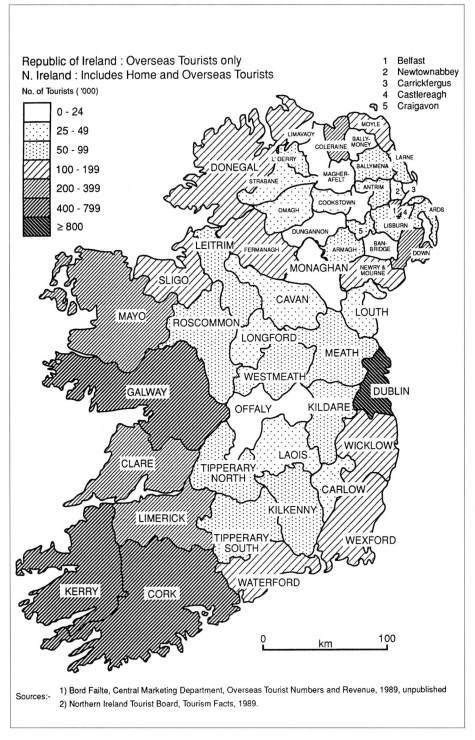

Sources:- 1) Bord Failte, Central Marketing Department, Overseas Tourist Numbers and Revenue, 1989, unpublished
2) Northern Ireland Tourist Board, Tourism Facts, 1989.

Fig. 1. The distribution of tourists in Ireland, 1989.

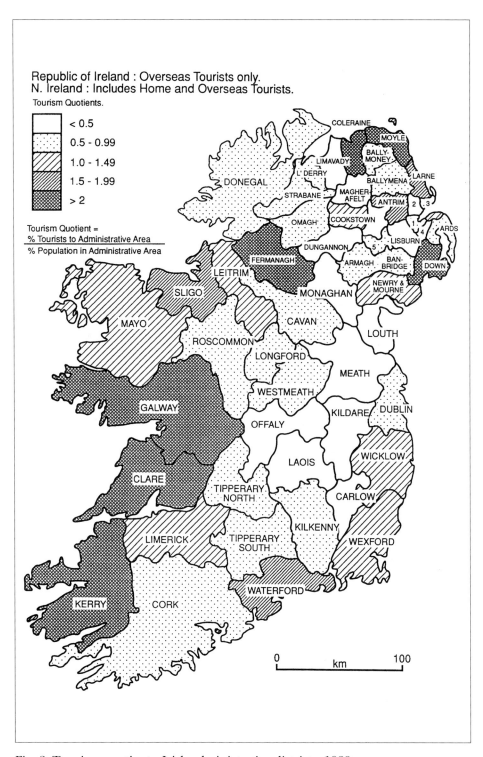

Fig. 2. Tourism quotients, Irish administrative districts, 1989.

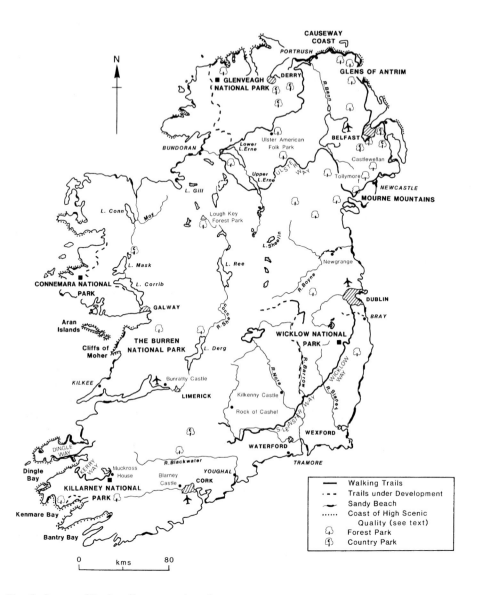

Fig. 3. Some of Ireland's recreational resources.

the drift-covered central lowlands except for specialist activities such as fishing/
cruising on the Shannon or visiting certain heritage sites such as Clonmacnoise.
A peak is reached in Kerry, which includes the romantic and picturesque
mountains and lakelands of Killarney. Clare and Galway offer similar physical
attractions, as indeed do the western extremities of Cork, including Bantry Bay,
and much of Donegal. Neither Cork nor Donegal stands out, however. In the
case of Cork, this reflects the size of the county, holiday visitors tending to
gravitate towards the relatively small southwestern periphery. Donegal, despite its
undoubted widespread landscape attractions, fails to attract its 'expected' total,

reflecting both its distance from points of entry, and limited approaches that do not involve a Northern Ireland border crossing. As such, it relies heavily on Northern Irish residents especially from County Derry, well accustomed to the border and all its security paraphernalia.

The east is much less significant from an overseas visitor standpoint, although it has something of an unrealised potential that has been emphasised in recent reports (Convery et al. 1989; Convery and Flanagan 1991; 1992). The natural comparative advantage of the west and especially the southwest from a tourist amenity point of view is apparent from Fig. 3. Highlighted are stretches of coast of outstanding scenic quality, as identified by the National Coastline Study (Brady et al. 1972). In the Irish Republic these exclusively favour the west coast, while the uplands of the western counties also contain three of Ireland's five National Parks, with the Burren's lower-lying but unique limestone scenery and ecology also recently similarly designated. Beaches and sea-angling facilities add to the attractions of the west coast, while golf packages have been strongly promoted to bring high-spending visitors into the southwest (Convery et al. 1989).

In the east, only Waterford, Wexford and Wicklow exceed 'expectations'. Waterford and Wicklow are helped in that respect by their sea links to Wales and France. Further north, the Wicklow Mountains are less renowned to foreign visitors than the mountains of western Ireland, but they comprise granitic and metamorphic uplands that vie scenically with the latter in one of the most extensive semi-natural areas of the country. In recognition of its wilderness quality and the need to conserve such resources in the east for scientific and recreation purposes, Ireland's fourth National Park was designated there in 1990.

To the north of the Wicklow Mountains, Dublin draws many holiday-makers, as befits a capital city with a host of cultural attractions, while historical resources of the highest quality — including the megalithic tomb at Newgrange (County Meath) and high crosses and a round tower at Monasterboice (County Louth) — provide stops on a fascinating journey through time in the Boyne Valley. In this latter area, the concentration of outstanding monuments and increased visitor numbers have encouraged plans for an Archaeological Park (Mitchell and Associates 1992).

North of the border, the lowland glacial landscape of drumlins and moraines in County Fermanagh, so waterlogged that lakes abound, provides one of the chief holiday regions. The Erne system, like the Shannon, has boating facilities and fishing opportunities that are unsurpassed in Europe, and most of the cruising holiday-makers and anglers visiting Northern Ireland are concentrated in that county. Apart from Fermanagh, the attractions are very much coastal ones, with Coleraine, Moyle and Down well to the fore in terms of both scenic resources and visitors from home and abroad. Coleraine's natural attractions are principally those of its holiday beaches at Castlerock, Portstewart and Portrush, literally backed in each case by top-class golf courses, while Moyle boasts the chief section of the world-famous Causeway Coast and contains part of the Glens of Antrim. Down possesses the coastal attractions of Newcastle and its region, as well as the northern part of the Mourne Mountains. Whereas no National Park exists in Northern Ireland, recognition of the amenity value and scientific importance of these landscapes is provided through their designation as Areas of Outstanding Natural Beauty (Fig. 3).

Recreation

Recent and widely-based recreational data for the local resident as opposed to the tourist are available for Northern Ireland only, although some more geographically restricted surveys of parts of the Republic have been undertaken, such as that for Dublin residents in 1975 (Mawhinney 1975), and by the Economic and Social Research Institute (ESRI) as part of the County Wicklow Tourism study (Convery *et al.* 1989).

In Northern Ireland, the Sports Council for Northern Ireland, the NITB and the Department of the Environment jointly commissioned a survey of the leisure day-trip activities of 4400 households throughout the period April 1990 to March 1991 (NITB *et al.* 1992). It was estimated that a total of 36.6 million trips (or 23 per person) were taken by Northern Irish residents over that period. The distribution of those trips between activities is described in Table 2, which shows

TABLE 2. Principal day-trip activities of Northern Ireland residents, 1990–91 (after NITB *et al.* 1992).

	Trips (million)	*% of all trips*
Visits		
Beach or seaside resort	5.71	15.6
Country or Forest Park	3.37	9.2
Town park	3.15	8.6
Sightseeing, picnicking, or sunbathing	2.49	6.8
Sporting event (as spectator)	1.68	4.6
Theatre or concert	1.06	2.9
Community event	1.06	2.9
Historic building or garden	0.92	2.5
Museum or heritage centre	0.81	2.2
Nature reserve	0.70	1.9
Walks		
Walk 2–5 miles	5.60	15.3
Rambling or hill-walking	0.81	2.2
Sports		
Golf ˙	2.38	6.5
Cycling	1.50	4.1
Swimming or other water-based activity	1.43	3.9
Fishing	1.17	3.2
Equestrian	0.44	1.2
Motor sports	0.40	1.1
Hunting or shooting	0.37	1.0
Mountaineering, climbing or orienteering	0.15	0.4
Other outdoor activities	1.43	3.9
Total Numbers	36.63	100.0

the importance of walking and seaside trips. Parks of one kind or another are well patronised for their relaxing and restful qualities, supported, no doubt, in the case of the Country Parks by free admission. General sight-seeing, which would include car-touring, attracts large numbers, while among the 'sports trips' a particularly high participation level is recorded for golf. In that case golf parallels the tourist activity pattern, and is relatively much more important for the Northern Irish than for United Kingdom holiday-makers as a whole (MAI Research 1992).

A number of the active pursuits — golf, fishing and walking in particular — are indeed attractive to day-trippers and tourists alike. Many other activities, principally those of the more passive kind incorporated in the 'visits' category, are almost exclusively the preserve of the local population. Chief among these are the beach or resort trips, with Portrush (Coleraine District) and Newcastle (Down District) the most favoured destinations, other urban and peri-urban attractions (particularly the town parks and Country Parks), together with the more rurally located Forest Parks. If the town parks, for which no records are kept, are excluded, the largest demand is experienced at the Crawfordsburn and Lagan Valley Country Parks, both on the doorstep of Belfast, where visitor numbers were estimated at 1.4 million in 1991 (NITB 1992b). Again within easy reach of Belfast, the Forest Parks of Castlewellan and Tollymore (Fig. 3), with their 300,000 visitors in 1991, highlight the importance of the forest as a recreational facility.

The overall geographical impact of the leisure and day-trip market largely confirms the holiday-making pattern (Fig. 4), the main exception being the stronger pull exerted by Belfast by virtue of its population concentration, and the short length of the trips, 57% of which were of less than 32km. The coastal and mountain areas to the south of the city are well patronised, most lying within an hour's drive of central Belfast. To the north, the Antrim coast is surprisingly less well favoured, despite containing (in Moyle) many of the Glens of Antrim and the unique attraction of the Giant's Causeway, which, with 350,000 visitors in 1991, accounted on its own for almost 1% of all leisure day-trips. Only Coleraine,with its beach resorts and related leisure opportunities, manages to overcome the distance factor in drawing people in considerable numbers from the main population centres of south Antrim and north Down.

Although there are no supporting state-wide Republic of Ireland data, there is no reason to believe that recreational pursuits differ greatly. The geographical impact necessarily changes, with the principal region of origin shifting to Greater Dublin, which contains approximately 30% of the state's population. Day-trip demand is therefore likely to cluster within an hour or two's drive of the capital. This supposition is supported by Mawhinney's (1975) study, which pinpointed visits to the coast as pre-eminent in popularity. Much of this demand is accommodated in Dublin Bay itself, but extends 65km south to Brittas Bay to take advantage not only of the open countryside but also of the good beach and cleaner bathing waters. All in all, approximately one-half of the Republic's day trips were thought in the 1970s to be taken along the stretch of coast between Carlingford and Arklow (Gillmor 1985).

A more recent survey of the Dublin leisure market by the ESRI indicates a continuity of the same leisure attractions (Table 3), although the more expensive pursuits of golf, sailing and equestrianism, and new activities such as

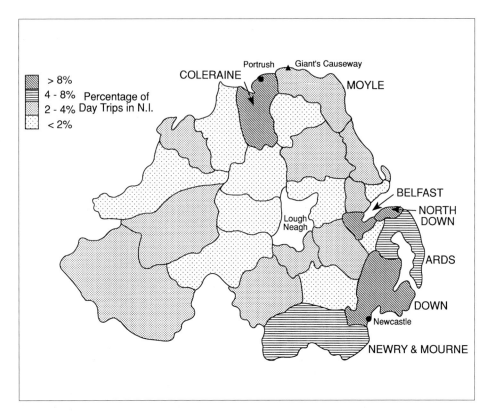

Fig. 4. Share of Northern Ireland's day-trip market by district council area, 1990–1.

TABLE 3. Recreational trips of Dublin residents with regard to County Wicklow (after Convery *et al.* 1989).

Activity	Trips (million)	% of trips to Wicklow
Beaches, swimming	1,946	13
Scenic drives, coach tours	1,741	34
Golf	1,227	9
Picnics	617	37
Horse-riding, pony trekking	606	29
Hill-walking, mountaineering	497	44
Historic places of interest	495	14
Freshwater fishing	172	22
Sea-fishing	116	22
Shooting	140	2
Orienteering	59	51
Coastal sailing, yachting	57	48
Inland sailing	54	38

orienteering, have no doubt increased their relative importance. While it is not possible to see the complete geographical pattern, County Wicklow stands out as a major playground for so many of the leisure pursuits of Dubliners, whether it be along its coastline or its mountain core.

To sum up this section, there is some degree of contrast within Ireland in tourist and day-trip activities, not so much between north and south as between western and eastern Ireland. Tourism, and particularly rural holiday-making, is reasonably well diffused, but with a definite western emphasis. In comparison, leisure day-trip recreation is population-based, and, with the great majority of the island's people living along the eastern seaboard counties between Antrim and Wexford, this inevitably means an eastern orientation. Combining the numbers involved, the greatest pressures deriving from tourists and recreational pursuits occur in the east, as does the greatest demand for access to the countryside.

Access to the countryside

Affording access to the countryside is not always a simple matter. Access to the countryside has, after all, been a matter of concern for the better part of this century in England and Wales, but, judging from the paucity of documentation in Ireland, public access has never been an issue to anything like the same extent. In Northern Ireland legislation for access has lagged behind that enacted in England and Wales: there, in response to demands building up in the pre-war years, the National Parks and Access to the Countryside Act (1949) required local authorities to map rights of way, and gave those authorities power to make agreements and orders or to compulsorily purchase land for the purpose of path creation. Meanwhile, the National Parks Commission took care of long-distance path provision. The Countryside Act (1968) also included provisions for wider access agreements as well as introducing Country Parks on boundaries of cities, so facilitating informal recreation for a highly urbanised population. Management agreements under the Wildlife and Countryside Act (1981) similarly allowed for public enjoyment of a variety of rural landscapes, in addition to promoting management in the interests of conservation and enhancement of natural beauty (Glyptis 1991).

In Northern Ireland the Ulster Countryside Committee was set up following the Amenity Lands Act (1965) and the Country Parks date from that time. However, it was not until the 1983 Access to the Countryside (NI) Order that alignment with England and Wales occurred in respect of the assertion and protection of rights of way (Devlin, 1989), and local authorities were enabled to create new public paths and long-distance routes. Grants paid through the Department of the Environment provide for assistance towards expenditure incurred by the district councils, 50% for path establishment and maintenance, 100% for long-distance paths, and 75% towards the appointment of a Rights of Way officer for a three-year period. Despite the financial inducements, the level of take-up has been fairly limited. The only long-distance footpath is the Ulster Way, which makes a complete circuit of approximately 800km virtually around the periphery of Northern Ireland (Fig. 3). The Ulster Way was encouraged by the Sports Council and largely pre-dates access legislation, being established from the 1970s, but still includes lengths of public road where no cross-country agreements exist.

Devlin (1989) questioned eighteen district councils to determine their response to access opportunities and obligations under the 1983 Order. That response was shown to be disappointingly apathetic (Table 4). There has been some improvement since her study, with seven Rights of Way officers now appointed. Path work has continued, but on a small scale, with budgetary and staff constraints paramount, although landowners can also be less than amenable to such work, as the survey indicated. Lack of public demand is, however, an underlying factor of crucial importance. If no one is pushing for access, and no votes are perceived in providing it, then clearly a low priority and little money will be given to this issue.

The explicit consideration of access to the countryside, and provision for it in legislation, is even more limited in the Republic of Ireland. It appears in three guises: first, in connection with the National Parks; secondly, through the setting up of walking routes; thirdly, through legislation in respect of rights of way.

There is no general Act governing National Parks as in England and Wales, although the first such park (centred on the present Killarney National Park) was defined by the Bourn Vincent Memorial Park Act (1932). This required the Commissioners of Public Works to "maintain and manage the Park as a National Park for the general purpose of the recreation and enjoyment of the public" (Office of Public Works 1990, 15). However, encouraging public access is only one of a number of objectives that emphasise conservation in keeping with the International Union for the Conservation of Nature and Natural Resources model and, where conservation and access come into conflict, it is conservation that takes precedence (Fadden 1990). Broadly speaking, however, reconciliation of conflicting pressures is being sought through the introduction of management plans that provide for zoning. The first of these plans (for Killarney) recognises four such zones — a Natural Zone, a Cultural Zone, an Intensive Management Zone, and a Resource Restoration Zone, with the Intensive Management Zone containing the highly used areas where non-conservation objectives are emphasised. Even this approach, however, cannot fully eradicate conflicting demands or reconcile opposing interests, as the present voluble controversy regarding the environmental impact of proposed visitor centres at the newest of

TABLE 4. District Council attitudes and responses to access provisions (replies to questionnaires to 18 district councils) (after Devlin 1989).

Rights-of-Way Officers: 4 appointments; non-appointments due to 'lack of work' and 'budget problems'.

Mapping and maintenance of pathways: 6 had done so or were in process of doing so; 'lack of staff' was principal reason for non-fulfilment of requirement.

Public demand for access: 6 stated no demand expressed; 6 considered demand existed to some degree (all in urbanised east).

Promotion of access for recreation: All considered this necessary; 2 considered there was little support from central government.

Relations between district councils and landowners: 4 'amenable', 2 'indifferent', 6 'difficult', 7 experienced 'some difficulty' in negotiating access the National Parks in the Wicklow Mountains and the Burren bears witness.

Access to the countryside in a wider context in the Irish Republic is officially promoted by COSPOIR (the National Sports Council), and particularly its Long Distance Walking Routes Committee, since its establishment in 1978 (Wilson 1989). These routes are shown in Fig 3, and comprise stretches over country roads, public land (particularly of the state forest company, Coillte) and private land, following agreements reached through local committees. For the most part, formal access legislation is not involved. The authority of planning bodies, namely the county councils, to create public rights of way and the responsibility to maintain them is contained in Sections 48 and 49 of the Local Government (Planning and Development) Act (1963), but there has been only one instance in the past ten years of use of either of those sections of the Act (Foras Forbartha 1987; Environmental Research Unit 1992).

All in all, little has been done in either territory, because it would seem that there is little demand to do anything (even from special-interest groups such as rambling clubs). Partly this is due to an attitude of 'let well alone'. Access has never been a significant problem, and there is apparently no great wish to upset landowners by demanding or asserting rights of access that have only rarely been denied in the past. Such a tolerant attitude must derive from the rural nature of Irish society, where access has traditionally been taken for granted along Mass paths, school paths, drove roads, butter roads, bog roads, etc. (Murphy, n.d.), and where the population density and degree of urbanisation are so much lower than in England and Wales that recreation resources are able to cope in general terms. Ireland is fortunate in the presence of open country and coast in abundance, not only in proportion to the overall population of the island but also in terms of the geographical distribution of those resources. In comparison with the conurbations of London and the Midlands of England, Dublin and Belfast are not only fairly small-scale but are also situated where attractive recreational coastlines, mountains and moorlands, containing either a National Park or Area of Outstanding Natural Beauty, are to be found in the immediate vicinity of the only two really built-up areas of the island. Nevertheless, the very presence of attractive open landscapes on the doorsteps of Ireland's main cities inevitably leads to pressures and environmental problems within confined areas, and it is to that issue that consideration must now be turned.

Recreation and the environment

Ireland boasts some of the most unspoiled landscapes of western Europe, with coastal, upland, wild moor and lake environments second to none thanks to its escape from the worst ravages of industrialisation, intensive agriculture, and population increase. Nevertheless, threats to those landscapes clearly exist, not least from recreation itself. This interaction between recreation and the environment is discussed here in relation to the coast, the chief recreational resource of the island.

As noted earlier, Ireland's urban pattern inevitably concentrates pressures along the east coast, and nowhere more so than in Dublin Bay, which offers an invaluable recreation resource for the million or so people who live in the Greater Dublin area. At its peak the Bay supports recreational demand equivalent to 5% of that population, participating in a wide range of pursuits of which beach activities and walking dominate (Table 5). Attitudes to the area, as elicited for the Recreation, Amenity and Wildlife Conservation Study section of the Dublin Bay Water Quality Management Plan (Environmental Research Unit 1991), showed greatest dissatisfaction with litter (particularly along Dollymount Strand) and pollution or 'smell' (particularly in the region of Bull Island and Clontarf). Over 55% of those involved in leisure pursuits described the water quality as less than acceptable, impressions confirmed by bacteriological counts that exceed National Limit Values under EC Directive 76/160/EEC. The Management Plan also detailed objections from sports clubs, with yachting enthusiasts, board-sailors and subaqua divers not surprisingly well to the fore. The problem arises from the traditional view of the sea as "a convenient dumping ground for waste" (Carter 1989), with sewage slicks and waste from industrial and agricultural sources creating substandard conditions. While moves to treat sewage and to dump macerated sewage sludge in more turbulent waters out to sea are taking place, there is still much to be done, both here and in other more developed and enclosed sections of the eastern coast, such as Belfast Lough. Fortunately, though, Irish waters are not generally polluted, and do not suffer the problems associated with many inshore waters of western Europe and the Mediterranean Sea in particular.

Apart from water quality, coastal environmental problems relate principally to dune destruction, beach sand loss and beach sand fouling. Water quality is, of course, an external factor as far as those engaged in leisure activities are concerned. However, with respect to the dunes, they are central to the problem. Dunes are the most vulnerable of natural recreation environments, suffering from trampling, vegetation destruction, and hence wind erosion. Many dune sites have been subject to conservation and management studies in order to limit the deterioration that might eventually lead to loss of the amenity value altogether,

TABLE 5. Recreational activities and participation rates in Dublin Bay (after Environmental Research Unit 1991).

Activity	No. of persons	No. of persons at particular time
Sailing	c. 15,000	2,000–2,500
Board-sailing	500–1,000	c. 200
Rowing	300–400	c. 200
Canoeing	200–300	c. 100
Water polo	250–300	c. 100
Sub-aqua	c. 500	c. 300
Water-skiing, jet-skiing	c. 250	c. 50
Angling	700–1,500	500–1,000
Motorboating and cruising	c. 30,000	c. 600
Beach activities	c. 250,000	14,500-21,500
Walking, jogging, sitting, sitting in cars, picnicking	c. 250,000	c. 3,300
Golf	c. 1,200	c. 150
Sand yachting	20–30	c. 20

as well as some fragile and rare plant and animal habitats. An Foras Forbartha's (1973) study of Brittas Bay was one of the earliest, and led to some remedial restoration work by the local authority, involving vegetation-planting, hydro-seeding and the construction of hard-surface parking areas (Convery *et al.* 1989). Bull Island has also come in for detailed study as an ecological resource, including dune complex, beach, salt-marsh, lagoonal sand and mud-flats, but one which has high amenity value given its proximity to the city (Jeffrey *et al.* 1977). There are pressures on dunes in Northern Ireland too, particularly on those beaches that provide easy access to vehicular traffic, as is common along the north coast (Benone, Castlerock, Portstewart), and at places such as Tyrella on the County Down coast.

The beaches of Ireland are undoubtedly one of its greatest natural assets and, to the casual visitor, seemingly indestructible. Although their loss is less noticeable than any dunescape, disappearance can and does occur. The principal causes of this are the removal of sand for agricultural and building purposes, and the building of shore defences which often produce effects that are entirely unanticipated. Carter provided the example of the destruction of the beach at Portballintrae, Co Antrim, which has taken place over a period of 90 years since a jetty was constructed at the end of the nineteenth century, and involved the loss of 100,000m^3 of sand (Carter *et al.* 1983). In similar vein, since the building of the Portrush seawall in 1963, the width of the beach has been reduced by 30m and the level by 1.5m. Engineering works in the Irish Republic, such as those at Bray, Youghal and Lahinch, can also be shown to have had adverse effects on this vital recreational resource.

Finally, the possibility of beach sand 'overloading' is pointed out by Jeffrey *et al.* (1977), with excessive recreational use, allied to antisocial dumping of litter, reducing the pore space and thus the beach's ability to act as a filter in which aerobic activity decomposes the organic material. If the process goes too far, then anaerobic conditions could result, forming black sulphurous zones that are both aesthetically objectionable and malodorous.

Management of recreation

Sidaway (1988), among others, has pointed out that "conflicts between recreation and conservation can be solved by sound management and planning". In this section some of the means suggested to resolve the seeming contradiction between environmental conservation and recreational access on the coast are considered in an Irish context.

Throughout Ireland management is in the hands of a wide range of landowners, both public and private, and in Northern Ireland there is also a large involvement of non-government organisations, of which the National Trust is the main player. The National Trust owns 29 properties, over half of which are seaside sites, making it the largest coastal landowner in Northern Ireland. A little over 800ha of that land is in Counties Antrim and Londonderry, counties that contain a number of the most attractive and popular mass recreational sites in the island (Fig. 3; Table 6). Chief of these is the Giant's Causeway, which was designated a UNESCO World Heritage Site in 1986, a National Nature Reserve in 1987 and an Area of Outstanding Natural Beauty in 1989. Most of the day-to-day management is provided by the National Trust, although the local district council (Moyle) owns

TABLE 6. Visitors to some National Trust coastal properties in Northern Ireland (after NITB 1992b).

Property	Numbers
Giant's Causeway	350,000*
Portstewart Strand	94,014**
Carrick-a-Rede Rope Bridge	44,904**
Murlough Nature Reserve (Co. Down)	275,000

* Figure applies to the visitor centre owned by Moyle District Council.

** Figures are substantial underestimates as they relate only to those arriving by car during the period when the entrance or carpark is attended.

the carpark and visitor centre. Inherent in this is some potential conflict between the district council's wish to promote the Causeway to the maximum and the National Trust's interest in long-term conservation of the rock formations, the bays and cliffs, and the flora and fauna. For the time being, however, the present visitor numbers are fairly well contained, as relatively few stray from the tarmac-surfaced path to the main rock formations. Study of photographs of the rocks themselves taken over a century shows some damage and removal of blocks suggestive of vandalism or careless treatment in the past, but such activity is now generally discouraged by patrols and education (Watson 1992).

More serious management problems occur beyond the main Causeway, even though only 10% of visitors (the more energetic) wander beyond that point. There fenced paths are established on unsurfaced laterite or weathered basalt. Fences on the lower cliff face are frequently damaged by rockfall, and those on the upper cliff path by animals or by climbing. At the upper level there is very little buffer between the cliff edge and the farmland, and thus a restricted natural habitat for the flora and fauna, so that visitors straying from the path cause damage, although probably less than nearby agricultural operations which change the habitat through high inorganic fertiliser inputs. Management is obviously necessary and expensive, involving three people working all year on path and related upkeep. However, there is some opposition to over-provision of high-standard paths and facilities, on the grounds that it is a so-called 'urbanising of the countryside' and in the belief that the experienced walker neither expects to be nor should be molly-coddled. The general aim of the National Trust is to ensure that by "maintaining a careful balance of low-key interference with geological features, restoration of path surfaces, drainage and limited fencing, the area is kept open and as safe as possible, without detracting from the wild and rugged Atlantic scenery that so many people come to enjoy" (Watson 1992, 43).

Some of the money required for essential maintenance work at the Causeway is diverted for this purpose from the National Trust property at Portstewart Strand, which is a net generator of income. This arises from the allowance of vehicular access, for which a charge is made during the peak season. Management problems relate to two separate areas, the beach itself and the backing dunes. On the beach there are conflicts between different recreational uses, including horse-riding and trotting, driving on the sands, and jet-skiing, apart from basic play and relaxation on the beach and in the water. Zonation of the property between vehicle and vehicle-free areas, and restriction of the inshore waters where jet-skiing is permitted, help resolve these conflicts, while equestrian pursuits are limited to the hours before 10.30 a.m.

A specific management plan is currently being developed with the aim of submitting the Strand for a 'Tidy Northern Ireland Seaside Award'. This requires attention to 29 criteria, including fulfilment of European Community standards relating to bathing water (Directive 76/160/EEC), industrial and sewage discharges, beach litter, emergency plans with respect to pollution incidents, easy and safe access, lifeguard, first aid and other safety requirements (National Trust 1992).

The dune area, in common with many similar environments in Ireland such as Dublin's Bull Island, is used for both formal and informal recreational purposes, with its ownership divided between the Portstewart Golf Club and the National Trust. Generally speaking, there is little public understanding of the fragility of dune systems, which are commonly but erroneously considered as permanent and resistant features (Eastwood and Carter 1981). Erosion damage arises chiefly from the informal pursuits of walking, jogging and children climbing and sliding. Fortunately from a management point of view, the majority of beach-users do not penetrate into the dunes. In that respect, the bringing of cars onto the beach can be seen as a positive advantage, compared with other dune sites where the access road and carparks lie behind the dunes, necessitating crossing the dunes to reach the sea shore. A proportion of visitors do, however, make their way into the dune complex, and some effort at containment is attempted through way-marking for walkers and joggers. Other activities, such as scrambling and horse-riding, have been banned since the National Trust took over in 1981. Eroded areas are made good by thatching and planting with marram grass, sometimes combined with dune profiling, techniques that are well tried and widely tested following guidelines advocated by, among others, the Countryside Commission.

Recent trends

Management strategies to cater for the use of recreational resources will become more and more essential as the demand for recreation grows inexorably. In Ireland that demand emanates from a number of sources, both spontaneously at the private individual level, and encouraged from within the public sector.

As far as the individual is concerned, the demand parallels trends throughout the western world in terms of increasing personal incomes and leisure time and, despite the high cost of motoring in the Irish Republic, a substantial growth in car ownership. What must also be taken into consideration, however, is demographic change and the additional recreational needs (both active and passive) of an expanding elderly population. In Ireland's case, the population of retired people could indeed grow faster than in many other European countries, as returning emigrants swell the numbers of pensioners.

Within the public sector there are particular groups that are increasingly encouraged to make use of the countryside and the recreational resources within it. The education sector is especially important in that respect, as school and further education curricula place more emphasis on practical studies involving fieldwork from the primary school up through the educational system. All parts of the countryside are liable to be affected by this trend, although the pressures will clearly be greatest closest to the main population centres. There will undoubtedly be positive advantages gained through the increased awareness, understanding and appreciation of the countryside and the fragility of some of its natural and physical systems. Such advantages can also be extended at the adult level through the greater provision of information at recreation sites, and interpretation centres at the more important locations. Some organisations positively advocate the expansion of their role in the educational field. This is above all true of the forestry services, where, apart from the links perceived between education, understanding, appreciation and conservation, there is a view that forestry with its high visibility in the landscape is often misunderstood and needs to explain itself to obtain more sympathetic acceptance (McCusker 1984).

At the local and national governmental levels there also exist the economic pressures to expand recreational facilities and opportunities, and particularly those that cater for tourists with their higher spending levels. Holiday tourism, with its greater concentration on the periphery (the west as far as the Republic is concerned, and the coast and Fermanagh lakelands in Northern Ireland) provides a natural means of redistributing income in the island. In national terms, tourism has a higher profile than in most other European countries outside the Mediterranean, contributing 6.8% of GNP in 1990 and 82,000 full-time job equivalents, even without taking account of any multiplier effects (Bord Fáilte 1991b). With its strong income and employment generating powers, tourism has been afforded a high priority, and its development is strongly supported by European Community Regional Development and Social Funds to the tune of IR£146.9 million out of the present IR£300 million programme, with the aim of strengthening this sector of the economy still further (Department of Tourism and Transport, n.d.). At present Ireland's status among European destinations is not high, for reasons that begin with its climatic and peripherality disadvantages and include to a debatable degree problems of prices, standard of facilities and perceptions of security. However, the positive response of visitors to the attractions of its landscape — its beauty, openness, freedom — and to the welcome from its people, together with a growing recognition of the need to produce strategies to promote and develop its potential at county as well as national level (Flanagan 1991), point to a substantial expansion in the near future. All the more will that be the case if peace can be obtained in the island, and if any shift occurs away from the sun, sand and sea holidays that have dominated the European tourism industry over the past 30 years, and towards the activity-type holidays for which Ireland caters.

Overall, trends point unequivocally towards increasing demand for recreation, for the provision of recreational land and for access to it. At present Ireland is undoubtedly lucky in its balance between population and recreational opportunities, although pressures clearly exist at the local level. That relationship between population, both resident and visiting, and the island's recreational resources does, however, need safeguarding through proper management and planning into the twenty-first century.

References

Bord Fáilte 1991a *The Product User Survey 1989/90*. Dublin.
Bord Fáilte 1991b *Tourism facts 1991*. Dublin.

Brady, Shipman and Martin, Niall Hyde (Consultants) 1972 *National Coastline Study*. Dublin. Bord Fáilte Éireann and An Foras Forbartha.

Carter, R.W.G. 1989 Resources and management of Irish coastal waters and adjacent coasts. In R.W.G. Carter and A.J. Parker (eds), *Ireland: a contemporary geographical perspective*, 393–419. London and New York. Routledge.

Carter, R.W.G., Lowry, P. and Shaw, J. 1983 An eighty year history of erosion in a small Irish bay. *Shore and Beach 52* (3), 34–7.

Convery, F. J. and Flanagan, S. 1991 *Tourism in County Meath — a strategy for the '90s*. Tourism Research Unit, University College Dublin.

Convery, F. J. and Flanagan, S. 1992 *Tourism in County Laois — a development strategy*. Tourism Research Unit, University College Dublin.

Convery, F.J., Flanagan, S. and Parker, A.J. 1989 *Tourism in County Wicklow — maximising its potential*. Tourism Research Unit, University College Dublin.

Coopers and Lybrand 1987 *Leisure Trips Survey*. Belfast. Coopers and Lybrand Marketing Services.

Department of Tourism and Transport (n.d.), *Ireland: Operational Programme for Tourism, 1989–1993*, Dublin.

Devlin, A. 1989 ". . . the countryside is theirs (the public) to preserve, to cherish, to enjoy and to make their own" (Lewis Silkin, 1948). Have policies for rural conservation allowed this to happen in Northern Ireland? Unpublished M.Sc. thesis, Queen's University, Belfast.

Eastwood, D.A. and Carter, R.W.G. 1981 The Irish dune consumer. *Journal of Leisure Research 13*, 273–81.

Environmental Research Unit 1991 *Recreation, amenity and wildlife conservation study*. Dublin Bay Water Quality Management Plan Technical Report No. 1. Dublin. ERU.

Environmental Research Unit 1992 *Planning statistics 1990*. Dublin. ERU.

Fadden, D. 1990 Management of upland recreation. Office of Public Works (unpublished).

Flanagan, S. 1991 The role of local authorities in developing a tourism strategy with specific reference to the Midlands East Region. *Irish Geography 24* (2), 136–43.

Foras Forbartha 1973 *Brittas Bay planning and conservation study*. Dublin. An Foras Forbartha.

Foras Forbartha 1987 *Planning statistics 1986*. Dublin. An Foras Forbartha.

Gillmor, D. A. 1985 *Economic activities in the Republic of Ireland: a geographical perspective*, 302–37. Dublin. Gill and Macmillan.

Glyptis, S. 1991 *Countryside recreation*. London. Longman/Institute of Leisure Management.

Jeffrey, D.W. *et al.* (eds) 1977 *North Bull Island Dublin Bay: a modern coastal natural history*. Dublin. Royal Dublin Society.

McCusker, P. 1984 A new interpretation of forest recreation management. *Irish Forestry* **41**, 59–65.

MAI Research 1992 *United Kingdom Travel Survey 1991*.

Mawhinney, K.A. 1975 *Outdoor recreational activities of Dublin*. Dublin. An Foras Forbartha.

Mitchell and Associates 1992 *Environmental Impact Statement. Park Centre Building for Boyne Valley Archaeological Park*, Dublin.

Murphy, J. (n.d.) Rights of way. COSPOIR, Dublin (unpublished).

National Trust 1992 Portstewart Strand, 1992 – proposed management (unpublished).

NITB 1992a *Tourism facts 1991*. Belfast. Northern Ireland Tourist Board.

NITB 1992b *Visitor Attraction Report, 1991*. Belfast. Northern Ireland Tourist Board.

NITB, Sports Council (NI) and Environment Service 1992 Northern Ireland Leisure Day Trips Survey. Preliminary report (unpublished).

Office of Public Works 1990 *Killarney National Park Management Plan*. Dublin. Stationery Office.

Pollard, J. 1989 Patterns in Irish tourism. In R.W.G. Carter and A.J. Parker (eds), *Ireland: a contemporary geographical perspective*, 301–30. London and New York. Routledge.

Sidaway, R.M. 1988 *Sport, recreation and nature conservation.* Sports Council. (Quoted in Glyptis 1991.)

Watson, P. 1992 *The Giant's Causeway — a remnant of chaos.* Belfast. HMSO.

Wilson, P. 1989 The expansion of long-distance walking routes in Ireland. *Irish Geography* **22** (1), 48–51.

In: A. Fenton and D.A. Gillmor (eds) 1994 *Rural land use on the Atlantic periphery of Europe: Scotland and Ireland*, 173–84. Dublin. Royal Irish Academy.

CONSERVATION AS A LAND USE IN SCOTLAND

Donald A. Davidson

Abstract: Conservation is a topic of considerable importance in Scotland given the country's diversity of flora, fauna, geology, geomorphology and landscape. A review of conservation is timely because of the accelerating pace of rural land-use change and the new opportunities offered by the formation in 1992 of Scottish Natural Heritage. This paper begins by outlining the framework of conservation before discussing such land designations as Sites of Special Scientific Interest, National Nature Reserves, Environmentally Sensitive Areas and Natural Heritage Areas. In the latter part of the paper, the need for a more integrated landscape approach to conservation is emphasised. An approach based on drainage basins is proposed.

Introduction

The importance of nature conservation and landscape protection in Scotland needs no emphasis. Scotland is endowed with a wide variety of landscapes, ranging from intensively farmed lowland areas such as the Lothians, the lower Tweed basin or Fife to the open subarctic plateaus of the Cairngorms or the bleak extensive bogs of Caithness known as the 'flow country'. The magnitude of landscape diversity in Scotland over short horizontal and vertical distances is outstanding. The Scottish coastline is of the order of 10,000km long, with only about 5% of its length occupied by continuous development. The blend of landscape and seascape is of particular note and gives areas such as Orkney, the outer Hebrides and the Highland sea-lochs their distinctive character.

A review of nature conservation as a land use in Scotland is timely for several reasons. Rural land use in Scotland is undergoing substantial change, with such themes as diversification, set-aside and afforestation. This accelerating pace of change has implications for nature conservation. In the past, development and conservation seemed opposing forces; now it is increasingly being realised that there can be economic benefits from the adoption of conservation strategies. A review of nature conservation is also timely given the formation in 1992 of Scottish Natural Heritage — an agency formed by the fusion of the Nature Conservancy Council for Scotland (NCCS) and the Countryside Commission for Scotland (CCS). This agency has the exciting potential to combine nature conservation with wider landscape management, including access for recreation.

173

This paper begins by reviewing the framework of nature conservation in Scotland and goes on to discuss the various types of designated areas. Frequent calls have been made for an integrated approach to nature conservation and the final part of the paper explores the mechanisms by which such integration can be achieved.

The framework of nature conservation

Powers of designation and protection are given in the UK in the National Parks and Access to the Countryside Act (1949), the Countryside Act 1968 (1967 and 1981 in Scotland), the Town and Country Planning Acts (1971), the Wildlife and Countryside Act (1981), the Wildlife and Countryside (Amendment) Act 1985 and the Natural Heritage (Scotland) Act (1991). The first of these provided for the designation and planning in England and Wales of National Parks and Areas of Outstanding Natural Beauty, the improvement of access in the countryside, including creation of access agreements, and the acquisition of nature reserves. The Countryside Commission for Scotland was established in 1967. The Nature Conservancy originally established in 1949 was split in 1973 into the Nature Conservancy Council and the Institute of Terrestrial Ecology. In 1991 the Nature Conservancy Council in England was renamed English Nature and the Countryside Commission has retained its title and remit except for Wales. In Wales the Nature Conservancy Council and the Countryside Commission have merged to form the Countryside Council for Wales. In Scotland, Scottish Natural Heritage (SNH) was established as an independent government agency and is responsible to the secretary of state for Scotland. The three broad aims of this agency are:

1. to conserve and enhance Scotland's natural heritage;
2. to improve opportunities for responsible public enjoyment and appreciation of the Scottish countryside;
3. to ensure that natural resources are used sustainably.

The Natural Heritage (Scotland) Act defines 'natural heritage' as including "the flora and fauna of Scotland, its geological and physiographical features, its natural beauty and amenity". Of particular note is the inclusion of the term 'sustainability' for the first time in UK legislation. This term is interpreted as corresponding to 'aesthetic sustainability', a phrase used by Mather (1992) to describe conservation of nature and landscape amenity. The formation of SNH is to be particularly welcomed because this agency has the responsibility for dealing in an integrated and sustainable way with landscape and recreation as well as nature conservation issues. An additional challenge for SNH will be to broaden its approach to sustainability by also incorporating land resource conservation. In a persuasive review of conservation in Highland Scotland, Mather (1992) argued that there is need for a renewed examination of the possible deterioration of hill land. The degradation of such land highlights the case for the additional adoption of a land resource approach to conservation. SNH is giving particular priority to environmental education as a way of encouraging a wider appreciation of Scotland's natural heritage.

SNH, in taking over the NCCS and the CCS, has responsibility for National Nature Reserves, Sites of Special Scientific Interest (SSSIs), National Scenic Areas

(NSAs), recreational designations such as Regional Parks and long-distance footpaths. SSSIs may be notified in addition under European Directives as Special Protection Areas (Wild Birds Directive) or Special Areas of Conservation (Flora and Fauna Habitats Directive); wetland SSSIs of international importance may also be listed under the Ramsar Convention. The 1981 Act covers protection of individual species (Part I), conservation of nature and landscape (Part II) in rural areas, and public rights of way (Part III). Part II includes provision for entering into agreements with landowners to provide site protection.

The approach to nature conservation since the formation of the Nature Conservancy in 1949 has been to identify and protect areas of scientific interest as representative of the remaining natural and semi-natural biological, geological and geomorphological areas in the country (Livingstone *et al.* 1990, 10). The prime function of the Nature Conservancy under the 1949 Act was to identify and establish by agreement, lease or purchase a series of National Nature Reserves for protecting the most important habitats and providing an opportunity for detailed scientific study. Also, a national network of SSSIs was introduced. The Wildlife and Countryside Act (1981) strengthened the tendency for wildlife to be protected in habitat islands (SSSIs). This approach works well with remnant fens or woods in lowland situations, but difficulties arise regarding selection of boundaries for many upland SSSIs.

Designated areas in Scotland

The extent of designated areas in Scotland is shown in Figs 1, 2 and 3. In Scotland there are 1319 SSSIs, covering 10.2% of the land area. In England there are 3536 SSSIs on 6.0% of the land, whilst Wales has 816 sites covering 9.3% of the country (Nature Conservancy Council 1991). Within Scotland, there is tremendous regional variation in the occurrence of SSSIs. In the Highland Region, for example, SSSIs cover 16%, with the figure rising to 27% for Badenoch and Strathspey District primarily because of the Cairngorms. In Scotland, the most extensive SSSIs are in the uplands and the coastal zones, though there are also some extensive SSSIs in the flow country of Sutherland and Caithness. The distribution of National Nature Reserves is also shown on Fig. 1; these reserves cover 1.4% of Scotland, compared to 0.4% for England and 0.6% for Wales. In Scotland, one National Nature Reserve (the Cairngorms) accounts for 22.7% of the total; one Marine Nature Reserve has been proposed in Scotland at Loch Sween, but this has not been achieved.

Nature conservation cannot be considered solely on a national basis and brief reference needs to be made to international designations in Scotland. There are seven Biosphere Reserves (a UNESCO designation) in Scotland, including St Kilda, Beinn Eighe and Rum. St Kilda is also included in the World Heritage Convention Listing. Ramsar sites are sites designated under the Convention on Wetlands of International Importance especially as Waterfowl Habitat. The Convention was adopted in Ramsar, Iran, in 1971 and there are 21 designated sites in Scotland, totalling 16,779ha. As already stated, Special Protection Areas are areas designated under a European Communities Council Directive on the Conservation of Wild Birds. There are 25 such sites in Scotland, covering 27,299ha, with the largest on Rum extending to 10,684ha. Figure 1 shows the distribution of Ramsar sites and SPAs; such sites are given protection by their SSSI designation.

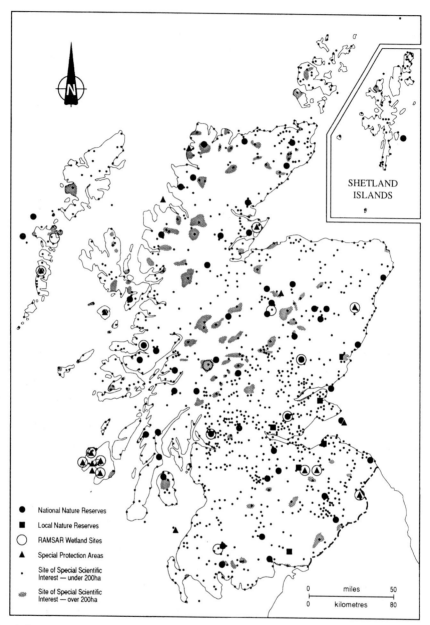

Fig. 1. Designated nature conservation areas in Scotland.

Sites of Special Scientific Interest

Sites of Special Scientific Interest have been established by the Nature Conservancy Council and its successor agencies, the NCCS and SNH. The sites are established on land considered to have special interest in terms of flora, fauna or geological or geomorphological features. Evaluative criteria used to select such sites are diversity, area, rarity, naturalness, representativeness, typicalness and fragility, and they are assessed with reference to the National Vegetation Classification Scheme. With proposed designation, any rural land-owner has to receive formal notification including a detailed map, statement of designation and a list of operations likely to damage the features of special interest. Management agreements are used as a means of reconciling conflict and they provide the basis for long-term site conservation. If a management agreement is made, then compensation is usually payable for loss of rights to improve land and convert it to a potentially more profitable land use. Compensation can be a lump sum or an annual payment, the latter applying only to tenants.

In a report dealing with assessing the effectiveness of management agreements, Livingstone *et al.* (1990) concluded that a list of potentially damaging operations can be confusing and daunting for owners and can prejudice the partnership approach which is being attempted. Overall, the report judged that the legislation has achieved its primary objective of preventing damaging operations from occurring. It was also concluded that the legislation is hardly conducive to a partnership — indeed, on the contrary, it provokes a situation of conflict. It is also important to appreciate that certain actions even outside the boundaries of sites can cause serious damage, and so planning authorities should exercise a strong presumption against any development which may affect sites (Selman 1992).

Land within SSSIs need not be limited solely to conservation; indeed, some mixed form of land use may well be beneficial to nature conservation. As an example, Dinnet Muir National Nature Reserve in Deeside has within its boundaries land managed for agriculture, grouse moors, recreation and archaeology. There may well be conflicts between differing nature conservation objectives. To take Dinnet Muir again, the reserve is outstanding for its birch regeneration as well as the distinctive suite of fluvioglacial landforms. With such woodland regeneration, the fluvioglacial landscape becomes increasingly difficult to see as an entity.

Environmentally Sensitive Areas

Agricultural over-production and intensification, as well as the threat of abandonment and pressures for more environmentally sensitive agriculture, have resulted in various schemes designed to support landscape conservation on agricultural land. There was the farm diversification grant scheme, now called the Farm and Conservation Grant Scheme, and in 1988 the Farm Woodland Scheme was introduced to encourage farmers to convert land to forestry, though the uptake has been minimal. With the set-aside scheme introduced in 1988, farmers receive payments if they set aside at least 20% of their land currently used for various surplus crops. No agricultural production is allowed on set-aside, though land must be kept in 'good heart'.

Perhaps the best-known scheme is the designation of Environmentally Sensitive Areas (ESAs). This scheme, administered in Scotland by the Scottish Office Agriculture and Fisheries Department, pays compensation to farmers who, on a voluntary basis within designated areas, manage land in accordance with conservation prescriptions. Thus farmers receive compensatory payments for agreements to follow specified farming practices or to carry out particular operations. The legislative framework for the ESA scheme is the Agriculture Act

(1986). The ESA scheme arose from an EC Regulation permitting member states to make payments to farmers in designated areas of high conservation value in order to encourage farming practices favourable to the environment. In Scotland the secretary of state has the power after consultation to designate ESAs if he or she considers that the maintenance of particular agricultural methods is likely to facilitate the conservation, enhancement or protection of natural beauty, nature conservation and buildings or other objects of historic or archaeological interest (Department of Agriculture and Fisheries for Scotland 1989).

The first two ESAs opened in 1987 (Breadalbane in Perthshire and Loch Lomond) and three more in 1988 (Machair of the Uists, Benbecula, Barra and Vatersay, the Stewartry area of Galloway, and the Whitlaw and Eildon area of the Borders). In 1992 five new ESAs were announced and two of the original ESAs were extended (Fig. 2) Conservation objectives were identified for each of these areas and formed the basis for drawing up standard requirements which are included in the agreements with farmers and crofters. For example, the Breadalbane ESA, located to the north of Perth, is an area of the highest scenic quality; two parts of the ESA are designated as National Scenic Areas. In this ESA, one conservation issue is to safeguard the existing low-intensity form of agriculture (extensive hill farming). Low profitability of sheep-farming leads to upland farms being sold for afforestation; there is also lack of expenditure on traditional farming operations which help to maintain/enhance the landscape. This can result in deterioration in the quality of semi-natural or broadleaved woodland and of the traditional field boundaries of stone dykes and hedges. Thus the agreed aims of the scheme for Breadalbane were as follows (Department of Agriculture and Fisheries for Scotland 1989, 13):

1. to protect the open hill rough grazing from land reclamation, over-grazing and the inappropriate use of herbicides and pesticides;
2. to provide similar protection for the unimproved, enclosed land in the valleys;
3. to rectify the neglect of traditional farm dykes and hedges;
4. to encourage natural regeneration of farm woodland;
5. to ensure that new developments such as vehicular tracks and farm buildings do not damage the landscape.

In Breadalbane there was also the recommendation that further forestry planting in the area should take account of the need to retain a balance between open hill and forestry. Applicants have to prepare a conservation plan, which is approved before the agreement is signed. The initial response within the Breadalbane ESA was very high; 75 agreements were made by April 1989, covering 31,260ha (total area of ESA c. 120,000ha) of which c. 90,000ha are estimated to be eligible agricultural land. The expectation is that between 60% and 70% of farmers and crofters in each ESA might ultimately join the scheme. Integral to the scheme is policy evaluation, which is achieved by environmental and socio-economic monitoring as well as surveys to check that agreements entered into within each scheme are carried out as agreed and to an acceptable standard. As an example, it is hoped that ecological and land cover monitoring will assist with judging the effectiveness of ESAs.

It is difficult to judge the success of the ESA scheme, since it has only been operational for a few years. In principle, the integration of agriculture with

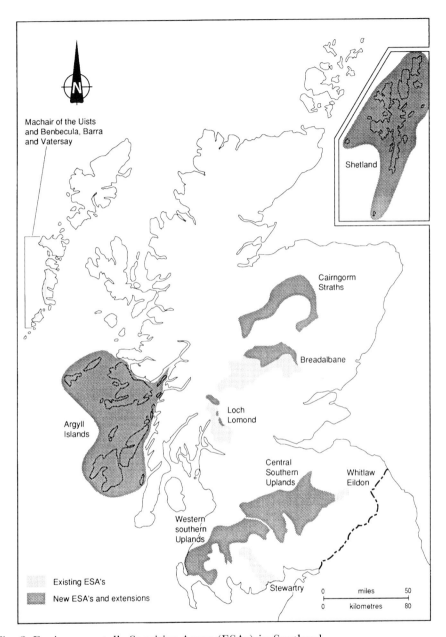

Fig. 2. Environmentally Sensitive Areas (ESAs) in Scotland.

conservation is to be commended. Mowle (1986, 17) considered that "the ESAs initiative amounts to little more than the introduction by the agriculture departments of their own system of site designation. This is likely to overlap with existing designations with concomitant confusion . . .". Such a view seems unduly negative, since the scheme does offer potential for a less sectoral approach to nature and landscape conservation.

National Parks and Natural Heritage Areas in Scotland

There has been a long-standing debate on the issue of National Parks in Scotland. The National Parks and Access to the Countryside Act (1949) made provision for such parks in England and Wales, but not in Scotland. The issue of National Parks in Scotland has been investigated by a range of committees; one of the earliest was the Addison Committee (1929–31), which recommended a park on the North American model for the Cairngorms. The most recent report proposing National Parks was published by the Countryside Commission for Scotland (1990). In essence this report proposed that four areas (the Cairngorms, Loch Lomond, Ben Nevis/Glen Coe/Black Mount and Wester Ross) should be designated as National Parks. The resultant administrative systems (planning boards or joint local authority committees) could thus arrange special management agreements so that these areas of high heritage value would be protected. Extensive consultation took place on this report (Countryside Commission for Scotland 1991), but the government rejected the proposals, considering them to be unnecessary and lacking in general support.

The government has introduced into the Natural Heritage (Scotland) Act (1991) a new countryside designation called Natural Heritage Area, which is described as having an ". . . outstanding value to the natural heritage of Scotland, and that special protection measures are appropriate for it . . ." (Subsection 6(1) of the Act). The main principle underlying the implementation of Natural Heritage Areas is the need for management statements which have been agreed by relevant individuals, agencies and authorities. The idea is for Natural Heritage Areas to subsume any existing NSAs within their boundaries, but to retain smaller areas of special conservation status (e.g. SSSIs, Ramsar sites, Special Protection Areas). The overall objective is to achieve integrated management on a voluntary basis by participating individuals, authorities and agencies. In a critical review of Natural Heritage Areas, Parnell (1992, 17) suggested that "with no extra funding, no dedicated staff on-site and no democratic authority in control, it is hardly conceivable that Natural Heritage Areas can achieve the level of comprehensive planning, management and investment that the present deteriorating condition of these areas demands". From 1 April 1992 it is the task of SNH to prepare proposals for the designation of Natural Heritage Areas.

Integration

From the 1950s to the 1980s a sectoral approach to rural land-use policy was very evident in Scotland, with the major land uses of agriculture, forestry and nature conservation following their almost independent paths. In a detailed analysis of the poor track record in achieving an integrated approach to rural land use, Mowle (1986) attributed blame to sectoral funding, with the different departments or agencies fighting for their own financial share with little or no reference to broader issues. He also criticised the traditional reductionist approach to problems, with scientists claiming expertise only in their own specialist areas. There are indications that during the 1990s more integrated approaches to the management of rural land use are becoming evident. The formation of SNH has already been described, with its remit for nature and landscape conservation, a major advance with the prospect of a more integrated approach. The ESA scheme offers better integration of agriculture and conservation. Another mechanism for encouraging integration between farming and conservation is through Farming and Wildlife Advisory Groups (FWAGs). The aim of such voluntary advisory bodies is to enhance the sensitivity of the farming community to conservation. Unfortunately FWAGs operate on very small budgets and the uptake rate by farmers has been low, at best 10% (Mowle 1986). Financial pressures have resulted in farmers and estate-owners seeking ways to diversify their enterprises, sometimes to the benefit of nature conservation. As an example, the rising interest in 'ecotourism' should encourage the adoption of a more integrated and sensitive approach to landscape conservation. In many ways this is not new in Scotland, since there is a tradition on some landed estates of an integrated approach to land management. Integrated use means that different land uses are managed in such a way as to achieve mutual benefit. Such an objective underlies the adoption of indicative forestry strategies which provide a framework for afforestation (MacKay, this volume).

Although there is evidence of more integrated and multi-purpose objectives in agriculture, nature conservation, urban fringe projects and forestry, there is still a dearth of such approaches at the national level in Scotland. It is the planning regions which seem most aware of the need for these approaches in dealing with strategic planning as exemplified by the Highland Region. As already indicated, a considerable part of this region is designated for nature or landscape conservation with 16% of its area covered by SSSIs and 22% by NSAs (Fig. 3). The special attraction of the Highlands is dependent upon areas of substantial scale being conserved for their natural landscape value. This means that it is important to maintain landscapes for 'open range' recreation. Therefore there are severe constraints on commercial afforestation. The Highland Regional Council considers that it has a major part to play in reconciling conservation and development interests. The aims of this Council are to develop and implement development policies to be compatible with (1) ecological processes, (2) preserving genetic diversity and (3) ensuring sustainable future use together with improving economic and social conditions. The objective is for conservation and development to be viewed in an integrated way. Thus the Highland Regional Council has a Conservation and Development Strategy designed to reconcile, in accordance with the World Conservation Strategy, the development of the region's economy with the need for conservation of its fragile ecology, and to promote the adoption of approaches to both land management and conservation that are not mutually exclusive. The Council has expressed concern over the number and size of SSSIs within its region and the fact that the location of some SSSI boundaries can pose particular problems to settlements and the provision of services.

One result of a sectoral approach to land management has been the proliferation of land designations, totalling about 45 (J. Miles, pers. comm.). The results of such multi-designations are particularly acute in such areas as Loch Lomond and the Cairngorms. In part, this difficulty is eased in the case of Loch Lomond, the presence of a Regional Park and its Regional Park Authority, the nearest equivalent to a National Park in Scotland. For the Cairngorms, a working party has produced a consultative report, which proposed a management strategy for protecting the natural heritage and ensuring economic and social benefit to local communities. The proposal was for such management to be achieved by voluntary cooperation; doubt has been expressed, for example by the Scottish

182

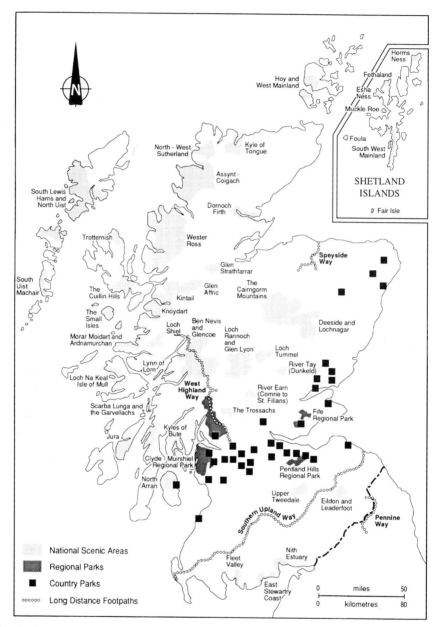

Fig. 3. National Scenic Areas, Regional and Country Parks and long-distance footpaths in Scotland.

Wildlife Trust, that such a voluntary mechanism would be effective. Instead, independent boards together with their staff are required to provide strong and effective management.

Although there are indications that better integration of land uses is being achieved within certain designated areas or management units, particular problems still arise regarding integration across ownership or designation boundaries. The key point is that areas of woodland, wetland or moorland do not occur as separate entities but are functionally linked with their surroundings. Linkage is achieved by the movement of water and its organic, ionic and sediment loads. Natural functional entities in the landscape are hydrological catchments, which ought to be the units within which integrated approaches to nature conservation are adopted. The need for a catchment approach to nature conservation in Scotland can be exemplified by brief discussion of the down-valley effects of soil erosion.

There is a growing awareness of soil erosion as a conservation issue in Scotland. Problems of footpath erosion in upland areas are well documented (Aitken 1986), but in recent years the incidence of soil erosion on agricultural land has been recognised. For example, Speirs and Frost (1987) relate the increase in soil erosion to the greater extent of autumn cereals. Fields prepared for spring sowing can also be highly susceptible to erosion should extreme rainfall events occur. This was the situation during the spring of 1992 in the Tweed basin in southeast Scotland. On 31 March 1992, an intense storm occurred with a maximum precipitation of 90mm in 24 hours. The result was that many fields prepared or recently sown experienced erosion along the tramlines formed by the agricultural machinery. Of course this was particularly the case with such tramlines following the maximum lines of slope. Typically these tramline gullies were 30cm deep and 40cm wide. In one severely eroded field on Loch Tower Farm 10km to the southeast of Kelso, 48 such tramlines were encountered in a survey along a 140m transect line; in the lower part of this field, the wash from these tramlines as well as from upper fields was concentrated into one large gully, which at its deepest measured 1.5m. A considerable amount of sediment was generated from this severe erosional event; material was deposited at the foot of the field, on the road near the gate and in a lower field. The farmer had to remove the eroded material from the vicinity of the gate and the resultant volume measured 73m^3. It is impossible to estimate the total eroded volume since much was lost from the field, but the eroded volume corresponding to approximately 110 tonnes indicates the magnitude of the erosional event. The on-site impact of such erosion is very obvious, with additional costs in field restoration by the farmer as well as lowering in crop yield. The off-site impacts can also be substantial, especially with respect to nature conservation. It is inevitable that some of the eroded soil was transported to streams during or after the storm event. There is thus the potential impact of such sediment on stream ecology. In this part of the Tweed basin there are a number of wetland areas, some designated as SSSIs. The erosion of the agricultural areas would also have had an impact on such habitats.

Conclusion

As already indicated, there is a clear swing in emphasis to an integrated approach within areas designated for some form of nature conservation. The example of ESAs with the integration of agriculture and conservation can be quoted; another example is the linkage of conservation and forestry as expressed in the seventeen Forest Nature Reserves in Scotland. Brief discussion of the soil erosion example demonstrates another aspect of land-use integration.

Environmental processes in one locality can well have impacts on habitats some distance away. Thus integration does not just mean a coordinated approach to land use within designated areas, but the implementation of such an approach at the landscape scale. The 'island' approach to designated areas or land uses needs to be replaced by a wider appreciation of the connectivity between landscape components. Scotland is favoured with many areas designated for nature conservation purposes, but the increasing challenge is to devise strategies for dealing with nature conservation in an integrated-landscape manner.

Acknowledgements

Thanks are expressed to the following, who commented on a draft of this paper: Les Firbank, Hilary Kirkpatrick, Mike Morris, Ian Simpson and Michael Usher.
Figure 3 is reproduced by permission of Scottish Natural Heritage.

References

Aitken, R. 1986 *Scottish mountain footpaths: a reconnaissance review of their condition.* Battleby, Perthshire. Countryside Commission for Scotland.

Countryside Commission for Scotland 1990 *The mountain areas of Scotland: conservation and management.* Battleby, Perthshire. Countryside Commission for Scotland.

Countryside Commission for Scotland 1991 *The mountain areas of Scotland: conservation and management. A report on public consultation.* Battleby, Perthshire. Countryside Commission for Scotland.

Department of Agriculture and Fisheries for Scotland 1989 *Environmentally sensitive areas in Scotland: a first report.* Battleby, Perthshire. Department of Agriculture and Fisheries for Scotland.

Livingstone, L., Rowan-Robinson, J. and Cunningham, R. 1990 *Management agreements for nature conservation in Scotland.* Department of Land Economy, University of Aberdeen.

Mather, A.S. 1992 Land use, physical sustainability and conservation in Highland Scotland. *Land Use Policy* **9**, 99–110.

Mowle, A. 1986 *Nature conservation in rural development. The need for new thinking about rural sector policies.* Focus on Nature Conservation Series, No. 18. Peterborough. Nature Conservancy Council.

Nature Conservancy Council 1991 *17th Report.* Peterborough. Nature Conservancy Council.

Parnell, B. 1992 Scots heritage agency faces new global challenges. *Planning* **969**, 16–17.

Selman, P.H. 1992 *Environmental planning.* London. Chapman.

Speirs, R.B. and Frost, C.A. 1987 Soil water erosion on arable land in the United Kingdom. *Research and Development in Agriculture* **4**, 1–11.

In: A. Fenton and D.A. Gillmor (eds) 1994 *Rural land use on the Atlantic periphery of Europe: Scotland and Ireland*, 185–94. Dublin. Royal Irish Academy.

CONSERVATION AS A LAND USE IN IRELAND

David Hickie

Abstract: The mechanisms and administration of conservation in Ireland are briefly described. Northern Ireland and the Republic of Ireland both have under-staffed and under-resourced government conservation agencies. Ireland has one of the smallest areas protected for conservation in Europe (~0.7% of the island). About 5% of the island is recognised as especially important for conservation; only one-tenth of this area is protected, while the remaining 90% is under constant attrition from a variety of developments, some of which are supported by public funds. The distinction is made between areas where conservation is the primary land use, and conservation in the wider countryside. Various means of expanding conservation as a land use are discussed, including improved screening of projects and economic programmes, the use of public funds to support benign development, and the restoration of degraded land.

Introduction

Since the island of Ireland is considered to be small, peripheral and agriculturally based, nature conservation has never assumed importance in official policy-making. Although Northern Ireland is part of the United Kingdom (UK), it has more in common with the Republic of Ireland as far as attitudes to conservation are concerned. Irish people have a low awareness of the need to conserve the natural environment compared with England or northern continental Europe, although Ireland has a higher environmental quality and more to conserve than its more populous and industrialised neighbours.

Conservation is becoming more prominent as an issue for three main reasons. Firstly, Irish people have begun to value their natural resources only when these are under threat. Secondly, Ireland's high-quality environment has been made a selling point for European tourists and Irish food. A third, related reason is the increased emphasis on environmental management in the reforms of the Common Agricultural Policy (CAP), and the social and economic changes taking place in the Irish countryside.

This paper deals with three topics: (1) mechanisms for conservation and its administration; (2) how conservation is put into practice; and (3) how progress in conservation can be made in the future. The paper is necessarily limited in

185

scope to discussing conservation *per se,* and the wider social and economic issues linked to conservation are only briefly mentioned here.

Mechanisms for conservation and its administration

Mechanisms for conservation

There are two main mechanisms for conserving areas of land for nature in the Republic of Ireland: Statutory Nature Reserves and National Parks. Nature Reserves are designated under the only piece of current legislation covering nature conservation — the Wildlife Act 1976. There is no legislation covering National Parks in a comprehensive manner, although this is expected in the near future. In Northern Ireland, National Nature Reserves and Areas of Special Scientific Interest are the main mechanisms, and these are designated under the Nature Conservation and Amenity Lands (Northern Ireland) Order 1985, which is intended to follow quite closely the British legislation (the Wildlife and Countryside Act 1981). There are no major differences between the operation of Nature Reserves in the Republic and in Northern Ireland.

National Parks in the Irish Republic are in line with International Union for the Conservation of Nature and Natural Resources (IUCN) standards, which require all land to be in state ownership and to be used primarily for nature conservation and public recreation and appreciation. There are no National Parks in Northern Ireland yet, in common with Scotland. If established, they would probably follow the unconventional English and Welsh model, where land is mostly not in state ownership but where special controls over planning and land management are exercised by the authorities.

Areas of Special Scientific Interest (ASSIs) in Northern Ireland are in parallel with Areas of Scientific Interest (ASIs) in the Republic. The major difference is that ASSIs have legal backing and the authorities have the power to control certain land uses even though the land may be privately owned. ASIs in the Republic of Ireland have no legal backing and the conservation authority has no power to control land use directly, although an amended Wildlife Act is expected to address this issue.

Scenic areas have received legal protection in Northern Ireland by means of designated Areas of Outstanding Natural Beauty. In the Republic, scenic areas can be designated by local authorities in their County Development Plan. Stricter planning controls can apply, and the designation has legal force by virtue of the County Development Plan being a legal document. A further mechanism exists in the Irish Republic — the Special Area Amenity Order — which has the potential to achieve landscape and habitat conservation, but this has been applied only to one small area to date.

Special Protection Areas (SPAs) are designated under the Birds Directive of the European Community (EC), applying to protection of the habitats of rare bird species in a European context. SPAs for habitats and species of animals and plants apart from birds will eventually be set up under the EC Habitats Directive. Environmentally Sensitive Area grant schemes (ESAs) have also been designated as a means of landscape and nature conservation in Northern Ireland, and are at a very early stage in the Republic.

Administration of conservation

The administration of nature conservation policies north and south is undertaken mainly by services within larger government departments, where conservation is only a small part of their briefs. Both services consist of very small numbers of dedicated and competent people who are overworked and under-resourced. The Environment Service in Northern Ireland operates under the Department of the Environment (DoE (NI)), while the National Parks and Wildlife Service operates under the Office of Public Works (OPW) in the Republic of Ireland. The Fisheries Boards are also conservation agencies, responsible for freshwater and inshore fisheries and pollution control, but they do not own or manage significant areas of land.

Independent advisory committees have been set up to advise the government on nature conservation policy. The Wildlife Advisory Committee in the Irish Republic has not met since 1986, but the National Heritage Council has been appointed to advise the government on the conservation of the built and natural environment and gives grants for conservation projects. The Council for Nature Conservation and the Countryside (CNCC) in Northern Ireland, set up in 1989, gives advice on conservation policy. The DoE (NI) allocates grants for conservation projects and also to voluntary conservation organisations.

In Northern Ireland, reserves are owned and managed for conservation by the National Trust for Northern Ireland, the Royal Society for the Protection of Birds, and the Ulster Wildlife Trust. In the Republic of Ireland, An Taisce, the Irish Wildbird Conservancy, the Irish Peatland Conservation Council and the Irish Wildlife Federation also own some small nature reserves.

How conservation is put into practice

Areas strictly protected for nature conservation by the state
About 5% of the Republic of Ireland is recognised as especially important for conservation. Just over half this area is terrestrial, the remainder being rivers, lakes, coastal bays and intertidal areas. Approximately 1600 individual ASIs have been identified to date in the Republic. All designated nature reserves and some areas within National Parks are included. Under 10% of all land identified as ASIs in the Irish Republic is used primarily for conservation (Temple Lang and Hickie 1992).

The total area of nature reserves in the whole of Ireland amounts to 21,734ha, with an additional 6670ha acquired in the Irish Republic but not yet designated (CNCC 1991; Temple Lang and Hickie 1992). The total area of National Parks is about 26,000ha. Not all of this area is of nature conservation importance, but all such land has outstanding landscape characteristics. The area covered by ASSIs in Northern Ireland is 6887ha, comprising 30 sites (CNCC 1991). It is assumed that even though such sites are in private ownership there is a reasonable chance of the conservation value of the land being maintained.

Therefore, a grand total of about 61,290ha is strictly protected for nature conservation, amounting to about 0.7% of the island of Ireland. In comparison, Scotland (which has a roughly similar land area) has 112,241ha designated as National Nature Reserves (Davidson, this volume). Many continental European countries have larger areas under strict protection than Ireland. Finland, for example, has about 2.6% of its area devoted to nature reserves and National Parka (Konturri 1986), while Hungary has 83,800ha or 0.9% of total area strictly protected (F. Márkus, pers. comm., 1991). The proportions of protected ASIs in the Republic (areas with the highest conservation value) under broad habitat categories are shown in Table 1.

TABLE 1. Estimated extent of protected Areas of Scientific Interest in the
Republic of Ireland in 1991 by habitat group.

Habitat group	Total area (ha)	No. of sites
Semi-natural woodland	3,732	30
Raised bog	2,153	14
Blanket bog	14,752	23
Fen	219	5
Islands (marine)	311	5
Sea-loughs	65	1
Sand-dunes, sand beach, shingle	2,358	4
Tidal rivers, estuaries	2,143	4
Freshwater lakes	2,501	4
Karst habitats	1,082	4
Grasslands	164	4
Others (mainly in Killarney Nat. Park)	4,440	
Estimated total	33,930*	

* To arrive at the estimated total area strictly protected for conservation in the whole of Ireland,
about 61,290ha, add the remainder of the area of National Parks, and the area under nature reserves
and ASSIs in Northern Ireland.

Sources: An Foras Forbartha 1981; Wildlife Service 1989; IPCC 1989; OPW staff, pers. comm., 1991.

Privately owned land of conservation interest

Areas of Scientific Interest in private ownership in the Republic of Ireland
have little protection beyond slightly stricter planning controls and refusal of EC-
funded capital grants for potentially damaging forestry and agricultural activities.
In practice, the refusal of afforestation grants on boglands of conservation
importance has effectively protected quite large areas of blanket bog (possibly in
the order of 50,000+ha). Some intertidal areas (100,000+ha?) are also effectively
protected because of their inaccessibility and physical unsuitability for
exploitation. Where most other habitats are concerned, the lack of any statutory
control or provision for management agreements in ASIs has been viewed as a
serious handicap.

Northern Ireland has 6887ha designated as ASSIs, where land is not owned by
the state but where damaging operations are controlled to some extent by the
DoE(NI) under management agreements with landowners, with provision for
financial compensation, and where the nature conservation value of the site has a
reasonable expectation of being maintained. The designation procedure for
ASSIs is very slow (Milton 1990) and there is a danger that sites could be
damaged before they receive formal designation.

International designations

The policy in both Northern Ireland and the Irish Republic is to designate EC
Special Protection Areas for those sites *already protected* in some form under
national legislation or which are not threatened in any way by exploitation. Of
the 20 SPAs designated in the Republic of Ireland, covering 5000ha, those that

are not already nature reserves are inaccessible cliffs or offshore islands. Only one SPA, a tiny island in Larne Lough, has been designated in Northern Ireland. A similar policy applies to putting the Ramsar Convention of 1971 into practice (where the objective is designation and wise use of wetlands of international importance): 21 Ramsar sites have been designated in the Republic of Ireland, covering 12,500ha, while only one — Lough Neagh/Lough Beg — exists in Northern Ireland, and most of these sites are already protected by the state.

Landscape conservation

The designation of Areas of Outstanding Natural Beauty in Northern Ireland is considered to be moderately restrictive in terms of planning controls on housing and other developments which are not considered to be sympathetic to the landscape. In the Republic of Ireland, local authorities appear to be less strict. National Parks in the Republic combine nature conservation and landscape protection in an integrated way, but these parks form less than 0.4% of the state. While Irish National Parks are expanding slowly, they are unlikely to occupy more than 1% of the territory in the foreseeable future. Landscape designations can partially protect some areas of nature conservation value by restricting certain non-exempted developments.

Environmentally Sensitive Areas

Environmentally Sensitive Areas also combine nature conservation and landscape protection objectives, but ESAs are environmental grant schemes for privately owned farmlands. ESAs have been operating in Northern Ireland since 1989, and include the Mourne Mountains and several of the Glens of Antrim, accounting for 40,000ha (Department of Agriculture NI, pers. comm.). In the Republic of Ireland, there has been very slow progress in accepting the concept of ESAs, and subsequently in designating areas. Several million pounds have been allocated for ESAs in Northern Ireland, but only £100,000 to date in the Republic. Only two small pilot areas have been designated there, in the Slieve Bloom Mountains and Slyne Head in County Galway, and no management agreements or payments have been made at the time of writing. Plans for designation of a sizeable part of County Fermanagh as an ESA are at an advanced stage, but may be delayed owing to UK regional financial cutbacks. The ESA schemes in Northern Ireland are probably the most encouraging aspect of conservation to date in the island as a whole.

The effects of development policies on the wider natural environment

Protection of the natural environment outside nature reserves and National Parks has been very difficult indeed to achieve. Little systematic information is available to date on loss or damage to ASIs in the Republic of Ireland. However, a recent assessment of 194 sites in four counties (Donegal, Sligo, Wicklow and Wexford) in the period 1983–91 suggests that the damage may be significant. An average of 37% of surveyed sites suffered some environmental damage and a further 16% were under immediate threat (Nairn 1992). While the sample is small, it highlights the urgent need for monitoring at national level and for legislative protection for ASIs. There have been some other significant changes.

(i) Corncrake *(Crex crex)* numbers have declined by 30% in the last decade. This decline is considered to be closely related to the great increase in silage-making and the concomitant reduction in haymaking as a land use (Mayes and Stowe 1989).

(ii) Of an original area of 311,000ha of raised bogs, only 17,970ha remain relatively intact, and no completely intact raised bogs now exist. About 94% of Irish raised bogs have been damaged or destroyed (IPCC 1992).

(iii) About 86% of blanket bogs have been damaged or destroyed, with 112,304ha surviving of national or international importance (IPCC 1992).

(iv) Over-grazing by sheep is now widespread throughout the western counties, leading to loss of heather and damage to peatland, sand-dunes and machair, and in places to soil erosion.

(v) A total of 23,412ha of commonage was divided in the Republic between 1982 and 1989, including 11,303ha in County Mayo alone (Land Commission, pers. comm.). Division of commonage has affected a number of important sand-dune and machair sites and upland areas owing to drainage and intensification, conifer forestry or golf course development.

Although the DoE(NI) and the OPW are charged with implementing conservation legislation, the main economic sectors and the government agencies which administer and promote economic development have a very great influence over, and responsibility for, the natural environment. These are considered briefly now with regard to specific sectors.

Agricultural policies. The industrialisation and intensification of farming, accelerated by EC agricultural policy over the last few decades, has resulted in the disappearance of many wildlife habitats and many farmers. The Department of Agriculture refuses grants for damaging farm developments in ASIs on the advice of the Wildlife Service in the Republic of Ireland, and this has saved some sites. Most arterial drainage projects had begun before such screening processes were set up, and since the mid-1980s no new drainage projects have been undertaken. No action has been taken by the Department of Agriculture to control over-grazing, which is being encouraged by EC and state policies. Commonage division is still officially supported. Action is being taken to control farmyard pollution, and considerable progress has been made by farmers since 1989, assisted by EC and state capital grants.

Forestry. Industrial forestry replaced agriculture as the main agent of land-use change in rural areas in Ireland in the late 1980s. The private sector has now overtaken the state forestry agencies in bare land planted per year. Total afforestation amounted to about 25,000ha in the Republic and 1200ha in Northern Ireland in 1991 (Forest Service NI 1991). The Irish Republic is now 6% forested, but 93% of the forested area is commercial conifer plantation with little conservation value. The remainder is composed of semi-natural broadleaved woods and old plantations, and those woods not protected are often threatened. Forestry policies have recently been directing afforestation towards wet mineral soils and enclosed marginal farmland. The Forest Service in the Republic has undertaken not to allocate EC-supported grants for afforestation of blanket bogs of conservation importance. The Forest Service in Northern Ireland is no longer acquiring blanket bogs, while Coillte (the Republic's state forestry enterprise) appears to have made a similar corporate decision.

Tourism. The expansion of tourism, now considered to be 'the great white hope' for rural areas, has both helped and hindered conservation. State acquisition of land for conservation is justified politically because of potential spin-offs for tourism. On the other hand, leisure developments such as golf courses have proliferated, damaging a number of important sand-dune systems, one of the most threatened habitat types in Ireland (Nairn 1990; An Taisce 1991). No assessment of golf courses as a land use has been carried out to date (A. A. Horner, pers. comm.), but the area under golf courses is probably expanding more rapidly than nature reserves.

Commercial peat exploitation. Commercial peat exploitation has affected large areas of raised bog, and Bord na Móna (Peat Development Board) machine-cutaway bogs cover about 80,000ha, mostly in the midlands. A sizeable but unquantified area is devoted to private peat-harvesting, which has reduced the area of raised and blanket bogs considerably. Again, state policy in both jurisdictions has been to encourage commercial peat production, but has largely ignored sites important for conservation. However, Bord na Móna in 1990 took a decision to sell 2500ha (20 sites) of unexploited bog to the National Parks and Wildlife Service (Bord na Móna 1990).

Other activities. Other activities that have reduced the area of natural and semi-natural habitat in Ireland are municipal rubbish dumps (e.g. in estuaries), roads, quarrying (e.g. geological sites and esker grassland and woodland), suburban development and aquaculture.

How progress in conservation can be made in the future

Since conservation as a land use covers only about 0.7% of the area of Ireland, and since it is the objective of conservationists to expand this area, how can progress be made in the future? In this section, it is necessary to make a distinction between nature reserves and National Parks on the one hand and conservation of the wider countryside on the other.

Expansion of strictly protected areas for conservation
The objective of nature reserves, National Parks and ASSIs is to protect a representative range of natural and semi-natural habitats and their individual plant and animal species. On present trends, the area strictly protected for conservation in reserves and parks will expand perhaps by about several hundred hectares or more each year (in comparison with the current acquisition of state forestry land in the Republic of Ireland of about 8000ha per year). But reserves and parks are not likely to cover any more than a tiny fraction of the island. The priority sites for future reserves are the most vulnerable and special sites. These are:

— sand-dune and machair systems,
— limestone dry grasslands,
— seasonally flooded grasslands,
— turloughs (seasonal lakes in limestone areas),
— raised and blanket bogs.

A certain amount of extra EC funds allocated under the Habitats Directive will be matched by state funds, and this should provide a boost to site acquisition in the next five years. The protection of existing sites is as important as the acquisition of new sites, since drainage of areas outside nature reserves can affect the protected areas themselves. While the continued acquisition of reserves certainly needs to be encouraged, it is only one part of a conservation strategy. The other part of such a strategy is to place a strong emphasis on conservation measures in the wider countryside.

Conservation in the wider countryside

There are a number of possible ways to achieve conservation objectives in the wider countryside.

Screening of development projects. Screening of development projects which are in receipt of state of EC funds needs to be improved, both at the level of state agencies and at EC level. Ultimately, environmental assessment of development programmes and policies will have to be applied, although this is being resisted by governments at present. An important element is openness in the state and EC bureaucracies, to allow for consultation with environmental agencies well in advance of published plans. If the screening process operates as it ought to, then there should be fewer land-use conflicts at local level.

Public funds for benign developments. A good example of using public money for the benefit of the natural environment is the EC-supported ESA scheme, pioneered in Britain and Northern Ireland. ESA schemes have generally worked well because they are simple, flexible and encourage farmers to farm in a particular way. The logical extension to the ESA concept is to replace conventional farm grants for activities which do not benefit the environment with annual premiums and capital grants for activities in tune with conservation, on a countrywide basis. This requires transfers of public funds from one budget line to another, since little additional money is available from the public purse.

The EC's Agri-Environmental Package accompanying the CAP reforms is considered very important in assisting conservation in the Irish Republic, since extra funding should be available for ESA-type schemes, long-term (twenty years) set-aside and extensification (reducing stocking levels). Over-grazing is an example of a phenomenon which could be reduced by imaginative use of direct payments, perhaps on a per hectare basis instead of per head of stock, to encourage farmers to reduce stocking while maintaining incomes, and by incentives to remove stock from hills in winter, including grants for sheep housing. This makes sense in the long term, since the existing payments have more of a social than a production objective.

The allocation of public money for forestry could have great benefits for conservation if more ecological principles could be built into the grant schemes rather than encouraging a type of industrial forestry with few, if any, environmental (or social) benefits. In general, a move away from large-scale infrastructural projects and more emphasis on smaller-scale, local development is likely to lead to environmental and social benefits.

Designations — how effective are they? Certain designations can be highly effective as mechanisms for conservation, such as nature reserves and National Parks which adhere to IUCN criteria. But such designations cannot be applied to large areas of the country. In Britain, designations such as SSSIs and National Parks are applied to relatively large areas, but recent surveys have shown that despite legal backing and provision for compensation, many SSSIs have been damaged, and conservation objectives have not necessarily been achieved in English and Welsh National Parks (Countryside Commission 1991). Ireland needs to learn the lessons of countries such as Scotland and England, which are more advanced in pursuing conservation policies. It is likely that ASIs in the Republic of Ireland will be given legal backing in the future, similar to the ASSI system in Northern Ireland. This will require more extensive survey work and more bureaucracy. A certain proportion of land can be conserved in this way, but Ireland should not have to rely solely on this strategy in order to expand conservation as a land use.

Restoration of land with conservation in mind. There is potential to increase the area of land of conservation value in state forestry plantations by allowance for open spaces within plantations and by management of certain areas in accordance with ecological principles. About 20–30%, or about 15,000ha, of Bord na Móna cutaway bog could be restored as wetland or scrub woodland. An advantage in both of the above cases is that the land is already in state ownership (An Taisce 1990; An Taisce and IWC 1990).

Expertise. Farming could not continue without the instruction of new entrants to farming. The same is true for forestry and fishing, and also for conservation. Conservation can be considered as an objective of all other uses of a renewable resource, because otherwise such land uses cannot continue in the long term. Conservation in nature reserves and in the wider countryside is vitally dependent on expertise gained through practical instruction, and, if progress is to be made, such instruction will need to be expanded.

Conclusions

(a) About 5% of the island of Ireland is recognised as specially important for conservation. Under one-tenth of this area is now officially protected.
(b) The area devoted primarily to nature conservation is among the smallest in Europe. Only about 0.7% of the island is protected, amounting to an estimated total area of 61,290ha. This compares with 112,241ha of Scotland occupied by National Nature Reserves alone.
(c) Areas of conservation value are continually being reduced by agricultural intensification, coniferous forestry, machine turf extraction, quarrying, roads, golf courses and other developments.
(d) The area reserved strictly for conservation is increasing annually, but on average by only a few hundred hectares per year.
(e) Land which can be regarded as effectively protected includes blanket bog where a veto on afforestation grants exists (~50,000+ha) and inaccessible intertidal areas and offshore islands (possibly 100,000+ha).
(f) The ability of the conservation authorities to enforce legislation and acquire or manage land in Northern Ireland and the Republic of Ireland is seriously hampered by lack of staff and finance.

194

(g) Some screening of damaging projects occurs, but it is insufficient to prevent the attrition of the natural resource base.

(h) The success of the ESA scheme in Northern Ireland, now covering 40,000ha and set to expand by three times this amount if sanctioned by the UK government, is the single most positive development in conservation in the island of Ireland to date.

(i) There is considerable potential for increasing the conservation value of farmland and providing support for farmers in the wider countryside if the EC Agri-Environmental Programme were implemented. This would mostly involve ESA-type schemes, with smaller amounts of land devoted to long-term (20-year) set-aside, and organic and less intensive farming methods.

(j) An increase in conservation as a land use in Ireland could be achieved on state forestry land, now occupying 5% of the island, and on cutaway bog owned by Bord na Móna, amounting to 80,000ha in the midlands, if even a small proportion of this land were restored and managed imaginatively in accordance with ecological principles.

References

An Foras Forbartha 1981 *National Heritage Inventory: Areas of Scientific Interest in Ireland*. Dublin. An Foras Forbartha.

An Taisce 1990 *Forestry in Ireland: policy and practice*. Dublin. An Taisce.

An Taisce and IWC 1990 Report to the Cutaway Bogs Committee appointed by the Minister for Energy (unpublished). An Taisce and Irish Wildbird Conservancy, Dublin.

An Taisce 1991 Impact of golf course development on the Castlegregory sand dune system, Co. Kerry (unpublished). An Taisce, Dublin.

Bord na Móna 1990 Conservation and corporate responsibility (press release).

CNCC 1991 Council for Nature Conservation and the Countryside. First Report 1989–1990.

Countryside Commission 1991 *Landscape change in National Parks* (CCP 359). Cheltenham. Countryside Commission Publications, UK.

Forest Service NI 1991 Annual Report of the Northern Ireland Forest Service. Belfast. HMSO.

IPCC 1989 Irish Peatland Conservation Council Action Plan 1989–1992. Dublin.

IPCC 1992 Irish Peatland Conservation Council Policy Statement and Action Plan 1992–1997. Dublin.

Konturri, O. 1986 Landscape ecology as a background to national nature conservation programmes in Finland. *Ale Fennica* 1 (1), 4–13.

Mayes, E. and Stowe, T. 1989 The status and distribution of corncrakes in Ireland, 1988. *Irish Birds* 4, 1–12.

Milton, K. 1990 *Our countryside our concern*. Belfast. Northern Ireland Environment Link.

Nairn, R. G. W. 1990 Fen lost in sand trap. *BBC Wildlife* (August 1990), 556.

Nairn, R. G. W. 1992 Areas of Scientific Interest: time for a rethink. *Ecos* 13 (2), 36–40.

Wildlife Service 1989 List of Areas of Scientific Interest in Ireland. Dublin. Office of Public Works.

Temple Lang, J. and Hickie, D. 1992 The Wildlife Act and European Community conservation measures — an up-to-date review. In J. Feehan (ed.), *Environment and development in Ireland*, 525–38. Environmental Institute, University College Dublin.

In: A. Fenton and D.A. Gillmor (eds) 1994 *Rural land use on the Atlantic periphery of Europe: Scotland and Ireland*, 195–208. Dublin. Royal Irish Academy.

LAND-USE PLANNING AND MANAGEMENT IN SCOTLAND

Derek Lyddon

Abstract: Eight recent policy statements bearing on rural land use are quoted to suggest a renewed need for a common approach and language if sectoral barriers and the gap between urban and rural planning systems are to be overcome. The national planning machinery suggested by a Select Committee in 1972 is compared with the current position and with the Dutch approach. The theory, practice and research connected with national planning guidelines are reviewed and related to new proposals for national planning policy guidelines. Some conclusions are set out concerning the purpose, process and product of planning at the national and regional levels.

"One of the most important results from the coordination effort is the realisation that tackling rural issues in a sectoral manner does not work" — the Scottish Office Rural Affairs Minister in *Rural framework* (Scottish Office 1991).

Introduction

During the last two years eight major policy documents which bear on rural land use in Scotland have been published. These include the white paper *This common inheritance: Britain's environmental strategy* (HMSO 1990), with its annual report on progress; a *Rural framework* from the Scottish Office (1991); a rural strategy from Scottish Enterprise (1991); forestry policy for Great Britain (Forestry Commission 1991); and the first operational plan from the Scottish National Heritage (1992), the new agency which combines the activities of the Countryside Commission for Scotland and the Nature Conservancy Council.

In addition there are now at least four committees or fora bringing together rural interests. These include: the rural policy coordination committee of all Scottish Office departments and the Forestry Commission, chaired by the under-secretary for planning in the Scottish Environment Department; meetings of departments, agencies and voluntary bodies in a Forum on the Environment; Rural Forum; and Rural Focus, a group of all the main agencies which the Scottish Office will chair.

All this indicates that a new opportunity exists to consider how policy coordination and integration can best be achieved. It also indicates that planning, as making provision for the future, and management, as the dovetailing of objectives, have never been more necessary in the field of spatial organisation and the environment.

This paper examines the experience of land-use planning and management in Scotland over the last twenty years, particularly at the national level, and makes some comparison with the processes adopted in the Netherlands. The aim is to see whether this past experience of the process and product of planning has a contribution to make to implementing these recent policy statements.

In all this a certain humility of approach is appropriate. The difficulties inherent in policy coordination and integration have to be acknowledged. Management by objectives and performance targets brooks no consultation. Coordination means persuading somebody to do something they had no intention of doing when they came into the room.

In addition, British administration seems to maintain a certain British reserve towards planning; this has been called an educated incapacity to see the need for it. The higher up the scales of planning one goes, and certainly at the regional and national scales, the more this seems to be true. The single-issue policy of the month, the initiative of the day, are preferred. Any coordinating strategy can be highly suspect; all too soon it will produce hostages to fortune. 'Integrated planning' has become politically suspect.

A common language and approach

Planning is noticed (and the planner blamed) only when something goes wrong. When it works it goes unnoticed as the natural order of things or market forces operating, so prompting the response: 'it would have happened anyway'.

Hence any consideration of land-use planning and management in Scotland may need to start with some consideration of what sort of planning is involved, and this is particularly true in relation to the Atlantic Periphery. Pending Maastricht and the proposed committee of the regions, which will have a particular emphasis on what has been termed 'spatial planning', there is already a directive on 'the assessment of the effects of certain public and private projects on the environment', and more recently a European Community (EC) programme of policy and action in relation to the environment and sustainable development.

The *Rural framework* report (Scottish Office 1991) proposed a new approach, which would aim to "develop a common language which rural communities and those working in their support can use to build their own ideas for development; and to bring together the considerable effort in support of rural communities in a continuing and coherent manner".

It was in 1979 that a UK countryside review committee and the UK conservation and development programme report *Putting trust in the countryside* (Countryside Commission 1979) both suggested the need for alternative ways of defining the problems and policy conflicts occurring in rural areas, seeking in fact a common language. The concern is with the essential characteristics of the rural economy and the land-use resource on the one hand, and the various forms of change which use that resource or affect the economy on the other — that is, the resource at risk and the changes at stake. Both conservation and development need to be redefined in the light of this.

The term development itself can be confusing because it can be taken to refer only to those changes in use or operations which are defined as development in the town and country planning legislation, with development being limited to those operations. Too often planning is then dismissed as a statutory procedure for planning and controlling urban types of development, with little relevance (or dangerous precedence) for changes in the rural economy and ecology. Before confronting this urban–rural divide, it is more relevant to conceive of planning as a basic human activity which is making arrangements or provision for the future, and land-use planning as making arrangements for the future change in the use of land.

Statutory town and country planning deals with only one type of change, which might now be called 'the built environment' but in the acts is termed 'development'. This nevertheless has a very wide span since it is defined as meaning "the carrying out of building, engineering, mining or other operations in, on, over or under the land or the making of any material change in use of any buildings or other land". This definition also applies in Ireland, and indeed from a review of an international manual of planning practice covering 26 counties (ISOCARP 1992) it can be observed how often this definition is followed in those countries which were not subject to the Napoleonic code. In all cases except the Netherlands agriculture and forestry operations "shall not be taken to involve development of land".

It might be said that rural planning and control of change is where urban planning was in the 1930s, with no statutory means of initiating and controlling change, and with the right of property-owners to do what they wished with their property. The built environment or development planning system set up in 1947 redressed this situation. It has five main components:

(1) a definition by Parliament of the type of change which is to be the subject of statutory planning;
(2) the duty of an authority to make arrangements for change (in for example a development plan);
(3) the need for an individual to seek consent to carry out change — with no compensation for refusal;
(4) the right to be consulted on a proposal, and the right for affected parties to object and to appeal;
(5) the power for central government to intervene in a decision of national significance.

Much will depend in future on the extent to which such a system could be extended in whole or in part to rural land use. Clearly the nature of rural 'property' is different insofar as it is the basic economic and ecological resource and source of livelihood, and the types of changes which can occur in the countryside are more varied and less easy to define.

Increasingly, however, rural property is being seen as common property. In place of the 'built environment' type of planning system, there is a wide range of subject legislation and single-issue agencies, with their own policy statements and, as Gilg (1992) has shown, a decade of countryside policies in some six subject areas. It is evident that EC and government policies and financial incentives which impinge on rural areas and land use stretch from social security through housing to tourism and transport, let alone agriculture, forestry and nature conservation; and all attempt to respect sustainability. The span of concern and variety of change are thus immense. Is it right, therefore, to suggest that any statutory system based on a duty to plan and control all types of rural change placed on a competent authority for a large area is now too late to achieve or not appropriate to the wide-ranging circumstances (not least harmonisation within Europe)?

As the distinction between town and country activities becomes more and more blurred, however, the dichotomy will become more evident. If, therefore, this imbalance between the two types of arranging for the future is to be redressed and if the inevitable sectoral barriers are to be overcome, past experience suggests that planning and management (not exclusively for rural affairs) will need to develop a *process* to which all concerned can contribute and a *product* which gives purpose to the traumas or *longueurs* of coordinating committees, and welds together the two types of control system.

The successes and failures on this front in Scotland over the last twenty years suggest that to achieve this process and product five essential criteria need to be met:

(1) an agreed record of land resource quality and significance, and the changes afoot: information as a corporate resource, as Coppock clearly indicates in this volume;
(2) a 'neutral' drafting team which all parties will trust;
(3) the capability to create a locational framework for multiple land-use priorities, impact assessment and resolution of conflict;
(4) a structured forum for advice and agreement on joint action, with an understanding of what is appropriate at each level of planning and management;
(5) the political will to move forward on all fronts.

Success clearly depends on moving on all five fronts together. While this has not been achieved in Scotland, the consideration given to the subject as a whole and the evolution of national planning guidelines may provide some pointers for the future.

National planning 'machinery'

It is evident from the five components of the British planning system referred to above that there is no duty on central government to do any planning. In the original Act, however, there was a duty "on the Minister of Town and Country Planning to secure consistency and continuity in the framing and execution of a national policy with respect to the use and development of land" (Cullingworth 1975).

Apart from the question of whether there can ever be a national policy rather than a bundle of policies (or a coherent set), the aim of securing consistency and continuity in framing and execution seems a useful description of what central government should be about. Shumacher (1973) put it another way: "it seems to me that the only intelligible meaning of the words 'a national plan' in a free society would be the fullest possible statement of intentions by all people wielding substantial economic power, such statements being collected and collated by some central agency. The very inconsistencies of such a composite plan might give valuable pointers."

The most important stimulus to planning at the all-Scotland level came from a report dealing with land resource use in Scotland (Select Committee on Scottish Affairs 1971–2). It is fruitful to review this committee's recommendations and the Government's response since, although twenty years old, they represent the last time the whole issue of the government machinery for planning of rural land use was fully ventilated and they are relevant to the current interest in this topic. The committee put forward five major recommendations: the formulation of national policy guidelines; a top-level working party to recommend the content of the policy guidelines; a commission on the urban environment; a Land Use Council drawn from individuals of high standing and long experience to act as a central forum for discussion of rural land use and to make recommendations to the government; and a Land Use Unit of professional planners (30 strong, equivalent to the number of planners already working in the Scottish Development Department (SDD)), to give advice to ministers on present land-use options and to warn of future difficulties.

The committee's recommendation that the government should draw up national policy guidelines was based on three arguments. If development plans by local authorities were to take into account social, economic and physical policies, guidance on what those policies were would be required. Examples of the topics they suggested should be covered in guidelines included major recreation, wilderness areas and the coast. The second line of argument arose from the committee's visit to the Netherlands. Here they found that the mutual physical planning relationships of the goals of national planning were illustrated in a map showing a structure for the whole country up to the year 2000. This form of coordinated national policy guidelines greatly helped to foster the team-work between central and local government. The third argument in which the committee found considerable merit and which bears on the question of Scottish policy guidelines arose from entry into the EC. It was in this context that the committee saw, even then, that Scotland will be a very modestly sized region, comparable in several respects to numerous other regions, many already having or in the process of producing strategies for development.

It was proposed that the Land Use Council should act as a central forum for discussion of rural land-use affairs. The members, who would be of high standing and long experience in their respective fields, should be appointed in their private capacities rather than as representatives of particular organisations. Among the topics which it was thought appropriate that the Land Use Council should discuss were: progress of experiments in the management of certain estates; proposed legislation on agricultural holdings affecting tenants; creation of protection forests; the transfer of land from agriculture to forestry; ways in which the Nature Conservancy Council and the Countryside Commission could play a more significant role; and the possibility of bringing together into one system the present varied arrangements to protect land which is of value by reason of its beauty, amenity and other special characteristics.

The committee attached great importance to planning machinery which is multi-disciplinary in countryside skills and where the staff do not have any prior loyalty to particular sectoral interests. They therefore recommended the setting up of a separate body of professional planners, about thirty strong, which they called the Land Use Unit. Illustrations of the functions which this unit would carry out were: to be a source of comprehensive understanding for ministers when agreeing on land-use policies and judging between apparently conflicting land uses, presenting options and warning of future difficulties; to provide an input to the national indicative plan; to assist in the provision of information requirements and to stimulate or guide research.

The government produced a white paper in response to the committee's recommendations (SDD 1973). They concluded that the main institutional innovations proposed by the committee would give rise to more problems than they would solve. It was thought that the shortcomings identified by the committee could be rectified by making better use of existing machinery and by making two significant additions. In the first place, however, the government stated that it was "their intention to intensify the efforts to give central guidance and to build up as quickly as possible a set of guidelines on those aspects of land use which should be examined for Scotland as a whole. They also intend as soon as this process permits to draw these guidelines together into a composite document which will be published. From time to time the document will be added to and modified as necessary. It will in time become a compendium of all that can usefully be said about the national framework for land use planning."

The first bit of new machinery which the white paper proposed was a Standing Conference of Regional Authorities meeting under ministerial chairmanship. One of its functions would be to consider from time to time how the proposed framework of planning guidelines could best be constructed and presented. The conference would draw upon advice not only from inside central and local government but also from bodies such as the Scottish Economic Council, the Nature Conservancy Council, the Countryside Commission, the Highlands and Islands Development Board, and indeed from outside experts. The Standing Committee was to be set up after the reorganisation of local government in 1975. In the event this did not happen.

The second piece of new machinery was a Standing Committee on Rural Land Use, with representatives of the six major agencies concerned with rural activities and environment. The committee would discuss land-use policies, and each representative could have an opportunity to ensure that decisions would take account of the various aspects of land use which were of major interest to it. The committee would also provide a forum for the discussion of the various topics which the Select Committee suggested would be suitable for consideration by a land-use council.

This committee was set up in December 1973 under the chairmanship of an under-secretary from the Department of Agriculture and Fisheries and lasted until the change of government in 1979. The committee played a vital role in the drafting of the national planning guidelines on rural conservation which were published in 1976 and which dealt, for the first time in one place, with land for rural conservation under the headings agriculture, forestry, nature conservation, landscape and the coast. If the committee had not taken part in reaching an understanding by all departments of the purposes of the guidelines and in providing a coordinated input, this first in the series would never have been launched.

There was no formal inter-sectoral liaison from 1980 until 1985 when the Scottish Office Departmental Group on the Countryside was set up under the chairmanship of SDD. This still exists and matches the composition of the original Standing Committee but appears much less formal.

National planning guidelines

By 1977 the provision of 'guidance' by government was beginning to be an idea in good currency. The government think-tank, the Central Policy Review Staff (1977), produced a report on the relations between central and local government. This stated that "a reasonable and moderate case can be made for establishing national guidelines on the basis that central government has the resources and perspective to enable it to form a balanced view on what constitutes a reasonable standard of service in normal conditions. Guidelines need not be rigid or mandatory; publishing them would imply that if local authorities want to ignore them they should be able to justify doing so."

A year earlier the Institute of Operational Research produced a report on health planning in Scotland (IOR 1976). It noted that "decentralisation should mean decisions being taken at the level which represents the best interests and values of the people affected by such decisions and at which relevant information can be obtained. In practice there will be a need for national guidelines within which local decision making can take place and therefore the level of decentralisation will be determined by the freedom these allow. Central involvement should be limited to those activities which cannot be devolved to a more local level and those in which the national interest is an overriding consideration. Guidance emanating from the central planning activities should not in general be prescriptive." The report also noted that even if the guidance or overview document contained little information which was new or unknown, this may not matter as long as the act of bringing it together on a common basis and in a form which was readily accessible and intelligible were to lead to more coordinated planning.

In both these quotations it is possible to note that guidelines were not seen as policy statements or as prescriptive, but as guidance which can be departed from provided there is reasoned justification.

The first opportunity to apply the Select Committee's approach came in 1973 when there was a 'client demand' for some national strategy to locate the many oil-related developments seeking coastal sites. A coastal planning framework was produced for consultation and confirmed by the publication of the *Coastal planning guidelines: North Sea oil and gas* (SDD 1974). As with all subsequent guidelines, the basis was a survey of the resources likely to be at risk and a review of the demand for them. In this instance the ecological, scientific and scenic characteristics of the coast were matched against the site requirements of a wide variety of oil-related developments.

In 1977 the National Planning Series was launched. The purpose was to develop a synopsis of the way the land of Scotland is used now and its potential for the future, and hence to put forward guidelines on some of the types of development which were likely to raise national issues. It was seen that such issues are likely to arise in two ways: from the particular characteristics of the land resource in certain areas which should be safeguarded in the national interest; and from particular activities such as industrial or recreational development having siting requirements which need to be satisfied in the national interest. It was also noted in the introduction to this first of the series "that planning guidance at the national level must steer between unhelpful generalisation and unwelcome direction. It should be based on a selection of those issues on which planning authorities and other agencies and developers might feel in need of guidance on where the national interest lies" (SDD 1977).

As the basis for this selection, land-use summary sheets were prepared. These take stock of a nationally significant land or spatial resource, assess the current

of the town and country planning system, its past and present performance and its proper role in the future" (Nuffield 1987). In the introduction to the report, Lord Flowers noted that they were not able to pursue in any detail the question of the European context of planning. They found, however, "that we need look no further than Scotland for some of the practices which we think could most usefully be adopted throughout the United Kingdom". The committee proposed a planning system based on (i) an annual white paper on land and environment, (ii) national planning guidelines at both national and regional levels, (iii) regional reports, and (iv) a county strategy, before getting to the development plan proper.

Their first recommendation was that "central government should publish from time to time concise and consistent statements of national policy wherever national interests in land use and development are at stake, and this should be its primary planning function". They suggested that the government should introduce national planning guidelines for England and Wales modelled on the Scottish system of national planning guidelines and prepared on a regional basis. The report noted that the guidelines would be brief and set out in simple language. They would go through a consultative procedure with local government before adoption. Much of what is currently in circulars could be put into guidelines, the committee suggested, releasing the circular proper for the task of giving advice on planning matters or matters of procedure.

The second major recommendation of the committee was that there should be a greater degree of consistency between land-use planning and other resource planning. It suggested the production of local strategies modelled on the Scottish regional report.

In response to the Nuffield Report, the Department of Land Economy at Aberdeen University carried out some research into the performance of the national planning guidelines, also funded by the Nuffield Foundation (Robinson and Lloyd 1990). In reviewing the characteristics and influences of the national planning guidelines they noted that "The essential characteristic of the guidelines has been the network of communications developed between SDD and planning authorities and between both these and other agencies in managing proposals for land use change. The approach has been one of conflict resolution through mediation and where possible consensus."

Robinson and Floyd made a number of qualifications, however, to the general positive contribution of the guidelines to planning in Scotland. "They have been selective, neglecting certain issues; there was confusion between circulars and guidance; some of the guidance has not been followed up (High Technology and Aggregates); some simply state the current position (Coast and Skiing); finally the strategic advice given has been carried through in an uneven way, in development plans."

A personal and pragmatic response to these points can be made. Zoning suggestions for aggregates were not followed up by the authorities, but an early warning of the superquarry at Glensanda on the shore of Loch Linnhe prepared the ground for approval without conflict. Countryside considerations and the skiing guideline did not avoid continuing argument at Cairngorm, but new facilities near Fort William suggested in the guidelines were opened up without creating a national planning issue. Finally, although the selection of high-quality sites for high-technology processes caused much argument and misunderstanding, firms found sites which had planning clearance and an option to purchase with immediate access.

In addition to this research and perhaps based on the central role which the Nuffield Report had in highlighting the value of national planning guidance as developed in Scotland, the then SDD commissioned the Planning Exchange in Glasgow to review all the planning documents put out by it. With regard to the guidelines, the research concluded that there was some uncertainty about their status in relation to other documents and that they were out of date; a long list of subjects which should be covered was suggested, and in any event a clear statement of national policy on all issues was needed.

In 1990 the government white paper, *This common inheritance* (HMSO 1990), stressed that the physical planning system is the centrepiece of policy to ensure that development and land use are compatible with proper care of the environment. The white paper gave an undertaking that the government would review national planning guidelines together with other planning advice to ensure that planning guidance takes full account of environmental considerations.

The Scottish Office *Consultation paper on the review of planning guidance* (SOEnvD 1991) proposed the introduction of a new series which would combine both policy and locational guidance, called National Planning Policy Guidelines, and the Department states that they will have significant weight in planning law. They should be regarded as statements of ministerial policy and will therefore be material considerations to be taken into account during the preparation of structure and local plans and in development control. Compared to the former national planning guidelines, the new guidelines should provide explicit statements of policy. Some will also contain a locational input (SOEnvD 1991).

As an example of the new type of document and the new style of presentation, the National Planning Policy Guideline on land for housing was issued for comments with a consultation paper, and one on business and industry has followed. The review is already well advanced on guidelines for skiing and mineral development. As a result of the consultation, the suggested order of priority to be given to additional subjects was: contaminated land; development in the country-side; coastal planning; green belts; landscape conservation; archaeology, sport and recreation. New areas and areas where the guidelines would benefit from revision included nature conservation, sustainable development, hazardous substances, forestry, and energy conservation. One of the results of this overall review and the Planning Exchange investigation, however, is that the land-use summary sheets will be discontinued, as "there is now a wide range of technical information of this kind available from other sources".

At regional level, the success of Strathclyde's Indicative Forestry Strategy, in its collaborative process of drafting and clarity of product, has been recognised. The Countryside Commission for Scotland's report *The mountain areas of Scotland* (CCS 1991) called for the preparation by regional authorities of indicative land-use strategies for forestry, agriculture, landscape and wildlife conservation, sport, recreation and tourism.

Conclusions

Purpose
Discussion at the symposium on rural land-use indicated that the purpose of

land-use planning and management at the national and regional levels needs to be more clearly defined. The experience in Scotland over the last twenty years and the current production of guidelines and indicative strategies suggest that some conclusions on the purpose can be reached. Some suggestions are set out in Table 2.

TABLE 2. Purpose of guidelines or indicative strategies for land resource use.

The purpose is to indicate:

WHICH resource uses or potential are of national or regional significance

WHERE such resources should be safeguarded, promoted or subject to environmental assessment

WHAT action should be taken by:
a planning authority in a development plan
developer in site selection
owner in site management
local level in further analysis

WHO will take decision on any proposed change in resource use

Overall,
an early-warning system
for
anticipating conflict
and
reducing uncertainty

Process and product

The present high level of interest in the planning and management of rural land uses in Scotland and the past experience of dealing with such issues at national level suggest that further coordination and integration will require an agreed process and product based on the following points.

(1) The pooling of information about land resource use as a corporate resource. This requires agreement on what facts to collect and monitor and who should pay.

(2) Resource assessments or summaries produced to a common format. Since the Land Use Summary Sheets are no longer to be produced by the Scottish Office Environment Department, a programme for alternative means of production is required. There could be a major role here for university departments working in conjunction with the resource agencies. Overall editorial control would need to be exercised, but this in itself would inspire cooperation, and the environmental education aspects should not be overlooked. The production of this basic survey material about nationally significant land resources surely cannot be left to chance.

(3) A strong but 'neutral' team for drafting assessments, advice and guidance. The number of people required has been consistently under-assessed. If consistency and continuity are to be achieved, such work cannot be contracted out.

(4) An agreement on the need for and content of guidelines and indicative strategies. There is evidence from several of the sources quoted that policy should be separated from guidance if a dialogue between national and local level is to be achieved in which the process of drafting is as important as the product. The more the guidelines are confined to policy which can be legally enforced, the less will they be able to coordinate actions which lie outside the town and country planning system.

(5) A structured forum or council for advice and agreement on joint action, with an understanding of what is appropriate at each level of planning.

It would seem that the political and administrative will is present to create a common approach, and to bring together the various 'statements of intentions' into what Balfour in this volume calls a framework of intent. Perhaps the time is ripe to review again the machinery of how this is to be achieved.

References

CCS 1991 *The mountain areas of Scotland*. Perth. Countryside Commission for Scotland.

Central Policy Review Staff 1977 *Relations between central government and local authorities*. London. HMSO.

Countryside Commission 1979 *Putting trust in the countryside*. London. Countryside Commission.

Cullingworth, J. B. 1975 *Environmental planning, vol. 1: Reconstruction and land use planning 1939–1947*. London. HMSO.

Forestry Commission 1991 *Forestry policy for Great Britain*. Edinburgh. Forestry Commission.

Gilg, A. 1991 *Countryside policies for the 1990s*. Oxford. CAB International.

HMSO 1990 *This common inheritance: Britain's environmental strategy*. London. HMSO.

IOR 1976 *Institute of Operational Research Programme of Studies in Health Planning*. London. Tavistock Institute.

ISOCARP 1992 *International manual of planning practice*. The Hague. International Society of City and Regional Planners.

Nuffield 1987 *Town and country planning. A report to the Nuffield Foundation*. London. Nuffield Foundation.

Robinson, J. R. and Lloyd, G. 1990 *National planning guidelines and the land development process: report to the Nuffield Foundation*. Department of Land Economy, University of Aberdeen.

Schumacher, E. F. 1973 *Small is beautiful*. London. Blond and Briggs.

Scottish Enterprise 1991 *Rural strategy*. Glasgow. Scottish Enterprise.

Scottish Natural Heritage 1992 *First operational plan*. Edinburgh. Scottish Natural Heritage.

Scottish Office 1991 *Rural framework*. Edinburgh. Scottish Office.

SDD 1973 *Land resource use in Scotland. The Government's observations on the Report of the Select Committee on Scottish Affairs*. Cmd 5428. London. HMSO.

SDD 1974 *Coastal planning guidelines: North Sea oil and gas*. Edinburgh. Scottish Office.

SDD 1977 *Large industrial sites and rural conservation*. Edinburgh. Scottish Office.

Select Committee on Scottish Affairs 1971–2 *Land resource use in Scotland, vol. 1. Report and Proceedings of the Committee*. London. HMSO.

SOEnvD 1991 *Consultation paper on the review of planning guidance.* Edinburgh. Scottish Office.

Witsen, J. 1977 Crucial physical planning decisions. *Planning and Development in the Netherlands* **9** (2), 102. Assen, The Netherlands. Van Gorcum.

While decisions are made by a central authority, each interest group communicates more or less exclusively with that authority and cross-sector communication tends to be minimised. Once diverse interest groups are brought together in a single structure, not as advisers but as policy-makers, then not only do they have to talk to each other, but also they have to work towards some kind of coordination, or *integration*, of their interests. This kind of integration (which social anthropologists refer to as 'cultural integration') is the process whereby diverse groups come to know, understand and possibly share each other's values. The imposition of an intermediary, in the form of a broker or a central authority, actually hinders this process by keeping the interest groups apart.

Participation and partnership

The distinction drawn here between consultation and participation is well understood in the debate over land use in Ireland. The concept of integrated land-use management often incorporates the idea that local communities should participate in decision-making (O'Hara and Commins 1991, 26); occasionally it is specified that the goals of development should be generated within local areas rather than being imposed from outside (Alexander *et al.*, n.d., 12).

Community participation in local development has been a prominent issue in the Republic of Ireland since the early 1960s, when rural communities, afraid for their viability under the government's economic policy, initiated schemes for generating employment and developing local resources (Tucker 1989). This movement had some success, for instance in the revitalisation of Gaeltacht areas (James 1988). But local initiatives could do little without state support, and although some state agencies have run schemes to encourage community enterprise, the government's overriding economic ideology was unfavourable to community development and many groups faded from the scene through lack of resources.

More recently, the principle of partnership between the state and local communities has been fostered by EC-funded schemes (Varley 1991). The major difference between this new trend and the earlier community development movement is that, with EC funding being, in many instances, more or less conditional on local community participation, the initiative comes now from the central authorities, what Varley has called "partnership from above" (*ibid.*, 84, 95).

A similar trend has developed in Northern Ireland, where, against the background of changes in the CAP, "the Government has accepted its responsibility to engage in and sponsor rural development" (Rural Community Network 1991, 2). A Rural Development Council has been created and is administering applications for funds and lobbying the EC on behalf of rural communities. The stimulus for this initiative has come, in part, from the Rural Action Project (RAP), a research programme funded by the EC and the Department of Health and Social Services. Between 1985 and 1989 the RAP examined the needs of rural dwellers in four selected areas and experimented with ways of using local resources to meet those needs (McConaghy 1988, 2). One such experiment represented an interesting reversal of the usual pattern of interaction between government and the community. A local development group in Belcoo, County Fermanagh, produced a draft Area Plan, which was then submitted to statutory agencies for comment (RAP 1989, 21–6).

Community participation and integrated land use

The question remains of whether community development initiatives, whether they spring from local action or state sponsorship, can lead to integrated land use. The primary objectives of such initiatives are to combat rural poverty and deprivation. It is inevitable, therefore, that local economic and social interests will predominate and, depending on the nature of the area in question, these may or may not incorporate a range of sectoral interests. There is a danger, however, that some sectoral interests, particularly those motivated by national and international considerations, might be ignored. Concern for the wider perspective in land-use management has prompted an argument which appears to run counter to the demand for local participation and initiative — namely, a call for greater central control of land use, through an integrated national policy (Gillmor 1979, 13, 20; 1989, 232).

National and international considerations underlie many environmental concerns. The principle that the global community has an interest in what is done locally is well established in green thought. It justifies people's concern for the Amazonian rainforest and for the future of elephants and rhinos in Africa; it motivates energy conservation policies and the search for more environmentally benign energy sources. In Ireland, for instance, international concern is expressed through Dutch involvement in the campaign to conserve Irish peatlands. However, there are many instances in which local social and economic interests are perceived as conflicting with wider concerns (for instance, see Milton 1990, 57), and this means that local community initiatives might not be sufficient to ensure that national and international interests are integrated into land-use management.

The importance of a partnership between local communities and the state is that it gives both a voice in land-use planning and creates the potential for local interests to be integrated with national ones. In order to ensure the integration of global interests into land-use management, the partnership needs to be extended to include representatives of those interests. For this reason it is significant that the SLMC, for example, includes the nominees not only of statutory bodies and local organisations but also of environmental groups such as the Royal Society for the Protection of Birds, the Ulster Wildlife Trust and Northern Ireland Environment Link, which have a wider constituency and close contacts with international organisations.

The prospects for integrated land use in Ireland

The preceding discussion raises several points concerning the prospects for integrated land use in Ireland. Firstly, examples such as the SLMC and the RAP Belcoo project demonstrate the potential for groups other than statutory bodies to participate in the policy process. Although the DoE(NI) has the legal responsibility to produce area plans and management proposals for designated areas, there is no reason why such plans should not be formulated by other groups and then adopted as statutory documents by the government. In other words, informal, voluntary co-operation between the state and other bodies can overcome apparent statutory restrictions.

Secondly, while a national land-use strategy would undoubtedly form a valuable basis for local programmes, the greatest potential for integrated land use lies in regions which are clearly definable and small enough to be handled as a single management unit. In Northern Ireland, the recent tendency has been to tie management plans to areas designated, usually for conservation purposes, by the DoE(NI), such as the Mournes, Strangford Lough and Lough Neagh (DoE(NI) 1992b).

Thirdly, the formal political structure and the prevailing political traditions and styles are clearly crucial in determining the feasibility of specific policies. The single-committee structure established for Strangford Lough received support across a wide range of local and sectoral interests. This level of unity almost certainly stems from the common experience, in Northern Ireland, of having to deal with central government over a wide range of issues.

It has been pointed out that, in the Republic of Ireland, the level of power and range of responsibilities delegated to local government is rather less than elsewhere in Europe (Barrington 1980, 39; Chubb 1982, 291). It is, nevertheless, significantly more than in Northern Ireland where, since 1972, district councils have been responsible for very few functions and often have a consultative rather than a decision-making role. In such circumstances, any management structure which appears to shift power away from the centre towards local and sectoral interests is likely to receive support from a wide range of groups, including local government, since all will perceive that they have something to gain. It is doubtful whether a structure of the kind established for Strangford Lough would be as widely popular where local government retains control of some important land-use functions and specifically, as in the Irish Republic, over planning control. The delegation of this power, even on an informal basis, to appointed, area-based bodies would also run counter to the personal, clientelistic nature of Irish politics.

Concluding remarks

The discussion presented here, rather than showing how integrated land-use management might be attained, has probably made it appear a less realistic prospect throughout most of Ireland. If so, then some of the difficulties involved in making it a reality should at least have been clarified. Because integration itself is interpreted in different and sometimes incompatible ways, there is a real danger of the parties involved in land-use management failing to communicate effectively with each other. There may also be good reasons why the prevailing distribution of interests and power relations obstructs the process of land-use reform. It is a truism, but one worth stating, that reform is only likely when it is perceived by those who wield power as being in their own interests.

This paper has been an exercise in comparative sociology. Ireland presents an ideal opportunity for this kind of study. The problems of land-use management, generated by the range of interests involved and the nature of the physical environment, are remarkably similar throughout the island. The social process of land-use management takes place within two different contexts, defined by contrasting state structures and planning systems. By examining the functioning of land use management in these different settings, it is possible to gain a better understanding of why certain policies and programmes succeed or fail, and thus to be better equipped to assess the prospects for land-use management in the future.

218

References

Alexander, D. *et al.* (n.d.) *The future of rural society.* Belfast. Northern Ireland Housing Executive.

Bannon, M.J. 1989 Development planning and the neglect of the critical regional dimension. In M.J. Bannon (ed.), *Planning: the Irish experience 1920–1988,* 122–57. Dublin. Wolfhound.

Barrington, T.J. 1980 *The Irish administrative system.* Dublin. Institute of Public Administration.

Brady Shipman Martin 1987 *Donegal–Leitrim–Sligo regional strategy.* Dublin. Brady Shipman Martin Urban and Regional Planning Consultants.

Brown, R.A. 1990 Bottom trawling in Strangford Lough: problems and policies. In C. ten Hallers-Tjabbes (ed.), *Proceedings of the 3rd North Sea Seminar 1989,* 117–27. Amsterdam. Werkgroep Noordzee.

Chubb, B. 1982 *The government and politics of Ireland* (2nd edn). London. Longman.

Convery, F.J. and Schmid, A.A. 1983 *Policy aspects of land-use planning in Ireland.* Economic and Social Research Institute, Broadsheet No. 22. Dublin.

Dawson, A. 1992 Changing farm economies in Northern Ireland: research report. *Anthropology Ireland* **2** (1) 16–19. Anthropological Association of Ireland.

DoE(NI) 1977 *Regional Physical Development Strategy 1975–1995.* Belfast. HMSO.

DoE(NI) 1989 *Mourne Area of Outstanding Natural Beauty: policies and proposals.* Belfast. Department of the Environment for Northern Ireland.

DoE(NI) 1991a *Strangford Lough: a consultation paper.* Belfast. Department of the Environment for Northern Ireland.

DoE(NI) 1991b *What kind of countryside do we want? Options for a new planning strategy for rural Northern Ireland.* Belfast. Department of the Environment for Northern Ireland.

DoE(NI) 1992a *Strangford Lough: management structure.* Belfast. Environment Service, Department of the Environment for Northern Ireland.

DoE(NI) 1992b *Management options for Lough Neagh and the Lower Bann.* Belfast. Environment Service, Department of the Environment for Northern Ireland.

Gillmor, D.A. (ed.) 1979 *Irish resources and land use.* Dublin. Institute of Public Administration.

Gillmor, D.A. 1989 Management of the countryside. In D.A. Gillmor (ed.), *The Irish countryside,* 226–35. Dublin. Wolfhound.

Healy, S.J. and Reynolds, B. 1991 Towards an integrated vision of rural Ireland. In B. Reynolds and S.J. Healy (eds), *Rural development policy: what future for rural Ireland?,* 41–67. Dublin. Conference of Major Religious Superiors (Ireland).

HMSO 1990 *Environmental issues in Northern Ireland.* House of Commons Environment Committee, First Report, HC Paper 39. London. HMSO.

James, C. 1988 Land-use planning for the Celtic languages. In R. Byron (ed.), *Public policy and the periphery: problems and prospects in marginal regions,* 26–39. International Society for the Study of Marginal Regions.

Komito, L. 1989 Dublin politics: symbolic dimensions of clientelism. In C. Curtin and T. Wilson (eds), *Ireland from below: social change and local communities,* 240–59. Galway University Press.

McConaghy, R. 1988 *Glens of Antrim Study Area Report.* Londonderry. Rural Action Project.

Milton, K. 1990 *Our countryside our concern: the policy and practice of conservation in Northern Ireland.* Belfast. Northern Ireland Environment Link.

Northern Ireland Office 1989 *Regional Development Plan for Northern Ireland 1989–1993.* Belfast. Northern Ireland Civil Service.

O'Hara, P. and Commins, P. 1991 Starts and stops in rural development: an overview of problems and policies. In B. Reynolds and S.J. Healy (eds), *Rural development policy: what future for Rural Ireland?,* 9–40. Dublin. Conference of Major Religious Superiors (Ireland).

RAP 1989 *Rural development – a challenge for the 1990s.* Londonderry. Rural Action Project.

Rural Community Network 1991 *Network News* (2). Cookstown. Rural Community Network.

Stationery Office 1989 *National Development Plan 1989–1993*. Dublin. Stationery Office.

Stokes Kennedy Crowley 1983 *A development strategy for the North East Region 1983–2001*. Dublin. Stokes Kennedy Crowley Management Consultants.

Tucker, V. 1989 State and community: a case study of Glencolumbcille. In C. Curtin and T. Wilson (eds), *Ireland from below: social change and local communities*, 283–300. Galway University Press.

Ulster Wildlife Trust 1991 *Strangford Lough: a consultation paper – response from the Ulster Wildlife Trust*. Crossgar. Ulster Wildlife Trust.

Varley, T. 1991 Power to the people? Community groups and rural revival in Ireland. In B. Reynolds and S.J. Healy (eds), *Rural development policy: what future for rural Ireland?*, 83–107. Dublin. Conference of Major Religious Superiors (Ireland).